EVERYTHING YOU NEED TO KNOW ABOUT EVERYTHING YOU NEED TO KNOW ABOUT INVENTIONS

Michael Heatley and Colin Salter

PORTICO

Published in the United Kingdom in 2012 by

Portico Books
10 Southcombe Street
London W14 0RA

An imprint of Anova Books Company Ltd

ISBN 978-1-907554-42-1

A CIP catalogue record for this book
is available from the British Library.

10 9 8 7 6 5 4 3 2 1

Designed by Blok Graphic, London
Printed and bound by Everbest Printing, China

This book can be ordered direct from the publisher at
www.anovabooks.com

"Doubt is the father of invention."

Galileo Galilei (1564-1642)

Contents_

0.3 Screw-cutting Lathe to Telegraph

0.4 Postage Stamp to Microphone

Contents_continued

0.7 **Photocopier to Laser**

0.8 **Manned Space Flight to the iPad and beyond**

Introduction_

No one knows who first declared that necessity is the mother of invention. Some say it was the Greek philosopher Plato in around 400 BC; others attribute it to a 17th-century English authority on fly-fishing called Richard Franck. But whoever it was, it is indeed need that drives mankind's genius for invention: the need for something more powerful, stronger, faster, more efficient than what was there before. Thus, for example, the steam engine replaces the water wheel, and is in turn superseded by the internal combustion engine.

Although we tend to characterize inventions as 'light bulb' moments, single flashes of brilliance, the reality is that most breakthroughs come after much trial and error. After that first spark of genius, a trail of heroic failures usually precedes the experiment that succeeds. The first successful typewriter came after over a hundred earlier attempts to solve the problems it presented. As Thomas Edison famously confessed, genius is 1 per cent inspiration, 99 per cent perspiration.

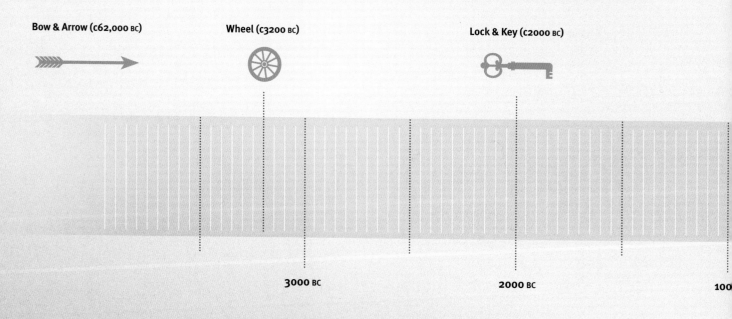

Bow & Arrow (c62,000 BC)   **Wheel (c3200 BC)**   **Lock & Key (c2000 BC)**

3000 BC   2000 BC   100

Where you see this symbol in the book it represents a leap forward in the development of an invention, in terms of both technology and timescale.

Even then, it's rarely the end of the story. New inventions are game-changers, raising new possibilities, showing us new horizons to be crossed and conquered. As Isaac Newton said, 'If I have seen further, it is by standing on the shoulders of giants.' Each invention builds on the technology and the applications of its illustrious predecessors, and each provides a new platform for further building. The invention of the telegraph depended on earlier inventions of the electromagnet and the relay, for example. The telegraph in turn led to the telephone, which gave rise to the transistor, on which all modern electronics depend.

And now it's your chance to build on the inventions of others. The book you are holding is a survey of 65,000 years of ingenious leaps of the imagination. Read the stories of inventions that changed and shaped the world we live in. Stand on the shoulders of the innovative giants between these covers and see where your imagination takes you.

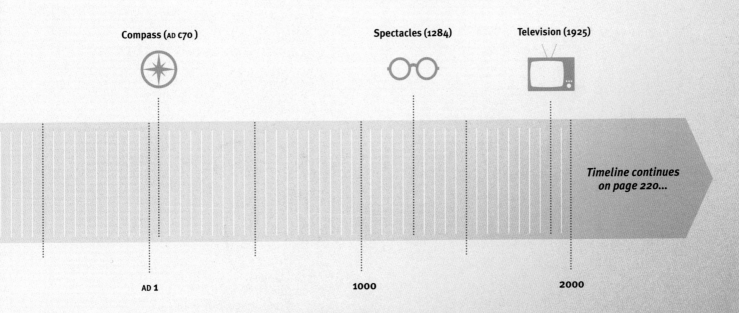

Compass (AD C70)

Spectacles (1284)

Television (1925)

Timeline continues on page 220...

AD **1**

1000

2000

Chapter 01.0

Bow & Arrow to Paper

c62,000 BC — AD 105

Bow & Arrow_c62,000 BC

Stone Age Man (Africa)

The bow and arrow represent mankind's first mechanical weapon, the first escalation in the arms race from merely throwing things – rocks and spears – at prey and enemies. The technology is simple and effective.

Developments over the millennia have been introduced to improve accuracy, range and speed of reloading; but the basic principle has remained virtually unchanged for about 64,000 years. One version of its latest incarnation, the underwater spear gun, uses exactly the same principle: firing a short spear by means of a powerful rubber band.

Something to Think About . . .

Otzi the Iceman, who was discovered in 1991 preserved in an Austrian glacier about 5,300 years after his death, was found with a 1.8m- (6ft-) long yew longbow and a quiver of 14 arrows, some with flint tips and feather flights. Was he a hunter? A warrior? Or a craftsman of bows and arrows? His end was not peaceful: he had head and hand injuries, and an arrowhead was lodged in his shoulder.

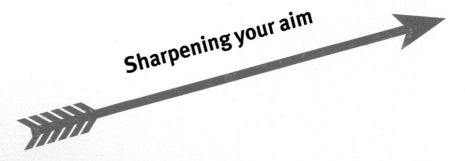

Sharpening your aim

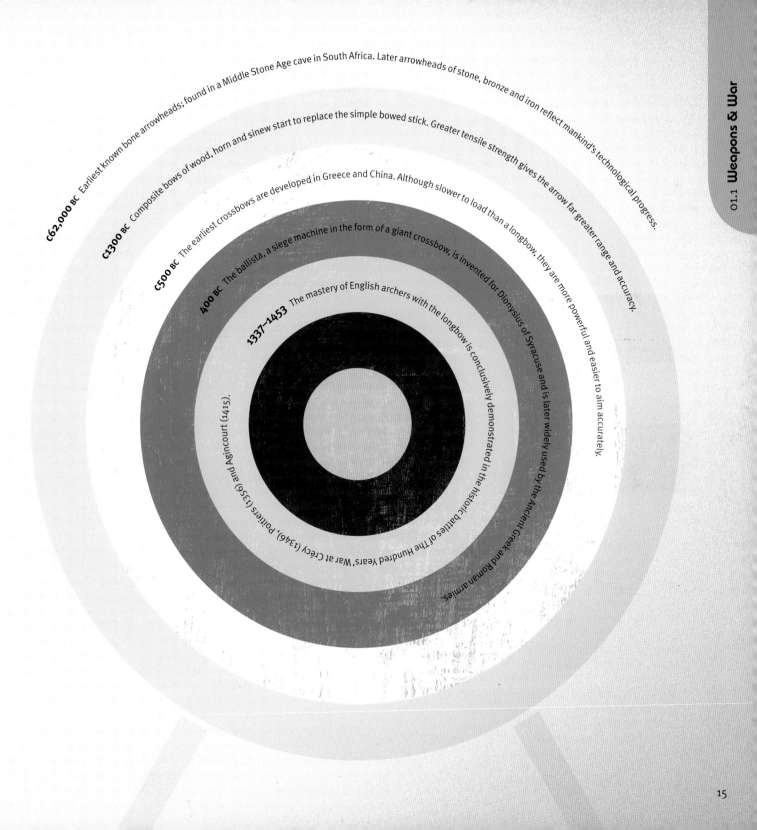

c62,000 BC Earliest known bone arrowheads; found in a Middle Stone Age cave in South Africa. Later arrowheads of stone, bronze and iron reflect mankind's technological progress.

c1300 BC Composite bows of wood, horn and sinew start to replace the simple bowed stick. Greater tensile strength gives the arrow far greater range and accuracy.

c500 BC The earliest crossbows are developed in Greece and China. Although slower to load than a longbow, they are more powerful and easier to aim accurately.

400 BC The ballista, a siege machine in the form of a giant crossbow, is invented for Dionysius of Syracuse and is later widely used by the Ancient Greek and Roman armies.

1337–1453 The mastery of English archers with the longbow is conclusively demonstrated in the historic battles of The Hundred Years' War at Crécy (1346), Poitiers (1356) and Agincourt (1415).

Grain Mill_c9500 BC

Nomadic Farmers (Syria)

The grain mill is a perfect illustration of necessity as the mother of invention. Corn has to be ground; by hand it's back-breaking work, so inventors have sought to harness available sources of power for the process. Only with the need for large-scale production was the millstone itself replaced – the new urban populations of the Industrial Revolution needed more flour than the small local mills could provide.

Since the 19th century steam and electricity have replaced wind and water in powering the machinery of mills, the word mill being applied to any powered building in which industrial manufacture took place. The inventions that first harnessed water and wind for grinding corn have since been applied to many other processes, including the supply of energy itself – for example in wind farms and water turbines.

c9500 BC
Grain is first ground in Syria, by moving one stone across another by hand, either back and forth (on a saddle quern or hand-mill) or in a circular motion (with a revolving quern).

Something to Think About ...

From the 6th century, some waterwheels harnessed the power not of running streams but of the ocean tides. The incoming seawater was trapped behind a dam, and then released to turn the wheel once the tide had ebbed. Evidence of these tidal mills survives in Ireland, France and elsewhere. In landlocked countries, Serbia for example, other water-mills made the most of fluctuating river levels by being housed on floating platforms or boats, anchored over the fastest currents of the stream.

Contents_continued

AD C850

The first grinding windmills appear in Afghanistan, rotating on a vertical axis. The more familiar horizontal axis first appears in Europe in the late 12th century.

C1860

The introduction of the faster, more efficient steel roller transforms the flour industry, and within decades the millstone, unchanged for 11 millennia, is all but obsolete.

C300 BC

In Greece the first horizontal waterwheels are developed and later geared vertical ones provide more power to turn larger millstones.

Plough_c6000 BC

The First Farmers (Mesopotamia)

Man's control over his environment began with farming. As soon as he discovered that breaking up the soil improved its productivity he began to invent new ways to make that process easier and more efficient. Each advance in plough technology increased crop yields and sustained ever-larger populations. But over-ploughing was also behind the environmental disaster of America's Dust Bowl in the 1930s. Ploughing techniques have also been applied to other activities such as snow clearance and the burying of pipelines.

c6000 BC

Something to Think About ...

The mouldboard plough made it easier to plough long, straight furrows, and so fields became long strips instead of squares. The length of such ploughed fields, about one eighth of a mile (200m), became known as a furrow-long and gave us the word furlong.

The scratch plough

In Mesopotamia tribespeople have the idea of harnessing domesticated cattle to a scratch plough, a version of the hoe they have been using to break the soil.

c1500 BC

c1000 BC

The crooked plough

Ancient Greeks take the simple wooden blade of the scratch plough and angle it forwards – crooked – to cut deeper into the soil, facing it in bronze or, later, iron.

The mouldboard plough

A curving mouldboard replaces the crooked blade, turning the earth over rather than just pushing it aside. This enriches the soil, and the furrows help drain the fields.

1730

The Rotherham plough

Joseph Foljambe of Rotherham devises a streamlined plough with an all-metal mouldboard. It is lighter, more manoeuvrable and therefore faster to use.

1784

The Scots plough

James Small of Dalkeith uses geometry to design a cheap, efficient one-piece cast-iron mouldboard. To make it widely available he refuses to register copyright.

The snowplough

Carl Frink of Clayton, New York, produces the first front-mounted snowplough. Prior to that snow was cleared by horses hauling a triangular wooden frame.

1920

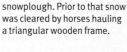

Cloth Dyes_c6000 BC

Unknown (Turkey)

The oldest known dye stuffs, from Anatolia, now modern-day Turkey, c6000 BC, were minerals such as red and yellow ochre. Organic sources, and ways of fixing them in the cloth, were sought from around the world and imported to Europe. In 1856 English chemist William Perkin invented the first synthetic dye, mauve, opening the floodgates for a vast range of new colours. In the 20th century man-made fabrics also required new ways to colour textiles.

Something to Think About . . .

The invention of the first synthetic dye was an accident, the result of experiments in synthesising quinine, a treatment for malaria.

Purple	
Indigo	
Blue	
Green	
Yellow	
Red	
Black	

Around c1800 BC the Minoan civilization is the first to make purple dye from murex shells. By 400 BC the shell is so scarce that in Byzantium purple is reserved for the Emperor and his immediate circle.	ONE	
India first produces deep blue dye and gives its name to the local plant from which it derived. It is exporting indigo to the Mediterranean during the first millennium BC.	TWO	
Woad is known as a source of blue dye in Neolithic times and Egyptian mummies are wrapped in woad-dyed cloth. Sadly the myth that Scottish warrior Picts painted their bodies in woad to scare the Romans is just that.	THREE	
Surprisingly rare as a plant dye. In medieval Lincoln in the east of England, green is made from a mixture of woad and weld. In the 18th century bright Saxon green is made from fustic (mulberry trees) and indigo.	FOUR	
Derived from weld (also known as Dyer's Rocket) since the Iron Age, and later from saffron. From the 18th century onwards quercitron (from oak) and fustic both give a brighter hue.	FIVE	
The madder plant provides a source of red dye in India from c2600 BC. In the 15th century the Mayans in central America make crimson from cochineal insects which is brought to Europe by the Spanish.	SIX	
In the 16th century Spain's conquest of South America brings logwood to Europe: a blue which becomes black when fixed ('mordanted') with ferrous sulphate.	SEVEN	

Weighing Scales_c5000 BC

Unknown (Egypt)

There are two types of measuring weight: relative and absolute. The former merely seeks to order two or more items by comparing their weight on a balance scale, while universally accepted measurements are necessary for the latter.

Many of the first and most enduring expressions of weight came from nature. In order to weigh small amounts precisely, the grains and seeds of plants were used. A grain of wheat became the grain of weight, while mustard seeds were used in India to weigh gold. The seeds of the carob tree gave us carats, still used today to express the value of gold and diamonds, although it has since become a metric carat. In Ancient Egypt the weights used were fashioned from bronze and often cast in the shape of animals; the cow represented one such unit of value.

Something to Think About ...

The Romans gave us the pound as a unit of weight. Pound is derived from the Roman word 'libra'. This explains why the pound is abbreviated as 'lb' and why the astrological sign Libra has an equal-armed balance scale as its symbol.

The earliest scales are found in an Egyptian grave. They are the precursor of the beam scale, with a horizontal bar on a pivot and two shell-like bowls at either end.

c5000 BC

2400 BC

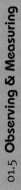

The first weights for use with balance scales are found in Pakistan's Indus River valley.

The bismar, a rod of wood with a large weight fixed at one end and a hook for the goods at the other, is the first recorded weighing device other than the balance scale.

Renowned German priest and clockmaker Philipp Matthäus Hahn builds a scale with a direct weight display.

British inventor R. W. Winfield develops the candlestick (spring) scale used for measuring letters and packages.

Experiments with electrical resistance by two American engineers leads to the digital scale, today's most common measuring device.

400 BC

1763

1840

1939

Wheel_c3200 BC

Unknown (Mesopotamia)

Without the wheel, nothing in today's society would move. But its inventor or inventors remain anonymous.

The wheel developed after logs were used to move heavy objects like sledges. The sledge created grooves in the logs that made them work more efficiently. The logs were then slimmed between the ends to take advantage of this, forming an integral axle with two wheels. Pegs were used to limit the movement of the axle creating carts, and when these were replaced with holes carved in the cart itself, the wheels and axle were made separately before being reassembled. Finally, the axle was connected to the cart frame for greater manoeuvrability, leaving the wheels alone to turn – a development reversed when engine-powered, rather than man-powered or horse-drawn vehicles, were introduced.

c3200 BC
Illustrations of chariots found in reliefs on the Standard of Ur in southern Mesopotamia, show the use of the wheel for transportation.

Something to Think About ...

The earliest known use of this essential invention was a potter's wheel that was used at Ur in Mesopotamia as early as 3500 BC. Wheels may therefore have had industrial or manufacturing applications before they were used on vehicles.

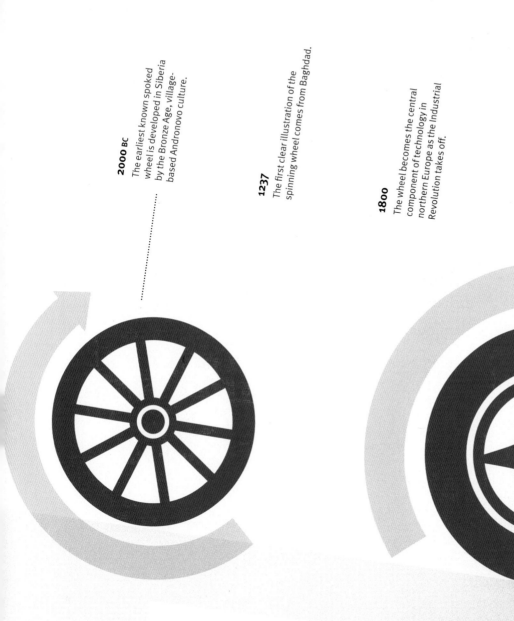

2000 BC
The earliest known spoked wheel is developed in Siberia based Andronovo culture.

1237
The first clear illustration of the spinning wheel comes from Baghdad.

1800
The wheel becomes the central component of technology in northern Europe as the Industrial Revolution takes off.

1887
The advent of the pneumatic tyre, invented by Scotsman John Boyd Dunlop, replaces the iron or steel rim, giving the wheel a new lease of life and many new transport-related applications.

As wheels have become increasingly sophisticated, so they can reach faster speeds. The current land speed record of 1,228km/h (763.035mph) was set in 1997 by Royal Air Force pilot Andy Green's ThrustSSC, while the piston-engined, wheel-driven record is 669km/h (415.896mph), set in 2008 by American Tom Burkland in his Burkland Streamliner.

Abacus_c3000 BC

Unknown (Asia Minor)

The abacus is a counting device with two sets of beads on parallel strings. The first set contains five beads on each string and allows counting from one to five, while the second set has two beads per string representing the numbers five and ten. The beads are counted by moving them up or down towards the horizontal beam at the centre. If you move them towards the beam, you count their value. A sharp movement along the horizontal axis moves all the beads away from the beam, resetting the device.

Something to Think About...

The abacus operates on a numerical base of five as humans first counted objects using their fingers. The use of two hands gave rise to the decimal system.

20,000 BC

Before
AD 1000

1764

1963

The Chinese suànpán –
a forerunner of the abacus –
uses hardwood beads on
a counting tray.

Pocket Calculator_1967

The pocket calculator is invented using
integrated circuit technology developed by
American physicist and electrical engineer
Jack Kilby of Texas Instruments. These were
soon seen in executive briefcases and, later,
the school classroom.

French mathematician Blaise
Pascal invents an adding
machine that can perform
additions and subtractions
directly and multiplication
and divisions by repetition.

Thomas de Colmar of France
produces the Arithmometer, also
known as the Thomas Machine. This
is reliably capable of performing the
four operations via a unidirectional
drum; division and subtraction
require setting a lever.

American inventor William Seward
Burroughs is granted the patent for
an adding machine with a full
keyboard and printing capabilities.
The first device prints only the
totals, but the individual entries
were printable by the time the
patent was granted.

Glass_c2500 BC

Unknown (Mesopotamia)

The first glass was almost certainly an accidental by-product of metalworking – the meeting of sand, soda and lime at a high temperature producing a substance of jewel-like translucence. Glassblowing techniques have changed little in over 2,000 years and glass has always been used for objects of expensive and delicate beauty.

In the wake of the Industrial Revolution mass production and science were applied, and modern glass is produced to precise specifications for its reaction to heat and light, its strength and flexibility.

c2500 BC

Glass beads
The oldest known glass artefacts, some decorative beads, are produced in Mesopotamia.

C100 BC

Blown glass
The art of blowing into a ball of molten glass to make a vessel is developed in Syria.

1st century AD

Cast glass
The Romans begin to use sheets of cast glass in the windows of fine buildings.

11th century

Broad glass
In Germany blown cylinders are cut open and flattened to make rectangles of glass.

14th century

Crown glass
In France glass is spun into circular sheets. The thick centre is called the crown.

15th century

Crystal glass
In Venice Angelo Barovier develops a pure crystal glass from sea plants and sand.

Something to Think About . . .

Industrial espionage is nothing new. Throughout the Middle Ages methods of production were closely protected – Venice banned foreign glassmakers from its factories, and its own craftsmen were threatened with death if they revealed their secrets.

1674

Lead crystal
George Ravenscroft of England produces lead crystal cut glass in imitation of Venetian glass.

1688

Plate glass
In France molten glass is rolled on tables and polished to make large sheets of glass.

1904

Mass production
American glassware manufacturer Michael Joseph Owens invents a machine for the mass production of glass jars and bottles.

1905

Vertical draw glass
The Fourcault Process pulls glass up vertically from a trough in very large sheets.

1959

Float glass
Sir Alasdair Pilkington (UK) makes huge sheets by floating molten glass on molten tin.

Iron Smelting_c2000 BC

Iron Age Man (South-western Asia)

From bells to cannons, from arrow tips to space rockets: since mankind first discovered iron in meteorites, we've used it to make everything from music to warfare. It is the most common element on the planet, and occurs in almost every product and process devised by the world's human inhabitants. Our inventive ingenuity has been as much applied to finding ways of extracting the raw material, as to making objects from it. Each new development of iron and steel production has represented a new industrial revolution.

Something to Think About...

It's the carbon content that makes the difference. Pure iron has none, but is too soft for most uses; workable wrought iron has 0.02–0.08 per cent carbon; flexible durable steel 0.2–1.5 per cent; and hard but brittle cast iron 3–4.5 per cent.

Ironing out a multitude of problems

1860s Germany

Karl Siemens' open-hearth process reaches higher temperatures by recycling hot exhaust and can make larger batches of steel to more precise chemical specifications.

1740s England

In Sheffield, Benjamin Huntsman invents a revolutionary process for manufacturing steel in clay crucibles, producing fine steel that can be moulded like cast iron.

1856 England

Sir Henry Bessemer devises a way of converting molten pig iron into steel in minutes by blasting compressed air through it. Cheap mass production is now possible.

c2000 BC Mesopotamia

The Hittites of eastern Turkey heat iron ore in a charcoal fire and hammer the results into worked or 'wrought' iron.

c1150 Northern Europe

The blast furnace, in which extra air is pumped into the fire, is capable of very high temperatures at which the iron absorbs more carbon and melts into moulds or casts.

1709 England

Abraham Darby of Coalbrookdale replaces charcoal with coke in his blast furnaces, and at a stroke lowers the cost of cast iron and starts the Industrial Revolution.

31

Lock & Key_c2000 BC

Unknown (Egypt)

Fastened vertically on the doorpost, the wooden lock contained moveable pin tumblers that dropped by gravity into openings in the crosspiece or bolt and locked the door. It was operated by a wooden key with pegs or prongs that raised the number of tumblers sufficiently to clear the bolt so that it could be pulled back. This method of locking was the forerunner of today's pin-tumbler locks.

American James Sargent invents the world's first successful key-changeable combination lock, adopted by the US Treasury. In 1873, Sargent patents a time-lock mechanism that is still used in bank vaults today.

Something to Think About . . .

The Ceremony of the Keys has taken place at the Tower of London for the past 450 years. Every night, the Chief Warder locks the Tower gates and, as he brings the keys back, is challenged by the sentry. He identifies himself as the bearer of 'Queen Elizabeth's keys' and is permitted to pass.

American inventor Linus Yale invents the pin-tumbler lock. His son improves upon his lock using a smaller, flat key with serrated edges, which provides the basis for modern locks.

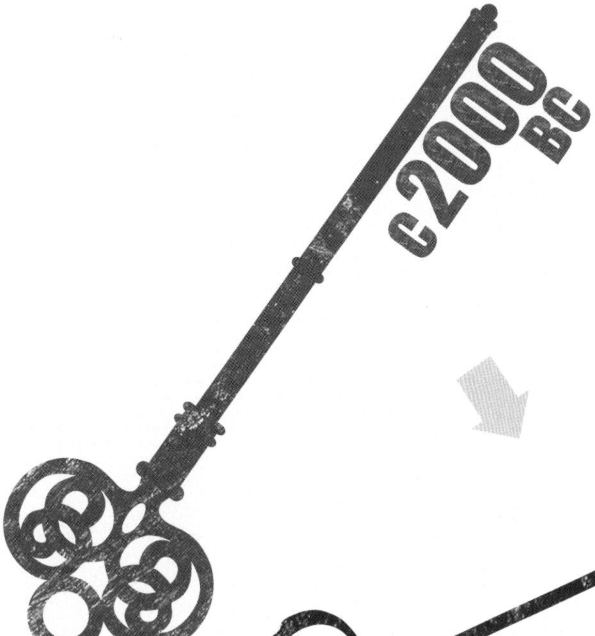

c2000 BC

Mechanical locks made of wood are in use in the Khorsabad palace near Nineveh, Egypt.

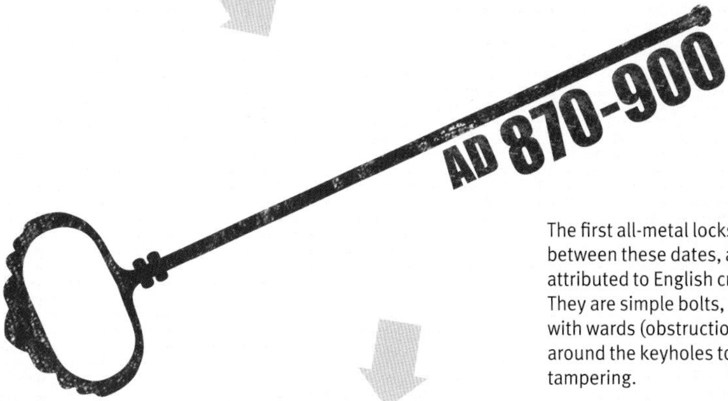

AD 870–900

The first all-metal locks appear between these dates, and are attributed to English craftsmen. They are simple bolts, made of iron with wards (obstructions) fitted around the keyholes to prevent tampering.

1778

Englishman Robert Barron patents a double-acting tumbler lock, a major refinement of the ward lock and still the basis of all lever locks today. The tumbler is a lever that falls into a slot in the bolt, preventing movement until the key raises it to the height of the slot. When each of the two tumblers is raised, the key slides the bolt.

1818

Jeremiah Chubb, a ship's outfitter and ironmonger in Portsmouth on England's south coast, introduces an improved four-lever tumbler lock.

33

Compass_AD c70

Unknown Geomancer (China)

Before the invention of the compass, finding your way around was a matter of using local knowledge, landmarks and the stars. Stuck in a strange country or on an unfamiliar sea, in fog, or on a cloudy night, you were lost. The compass liberated explorers on land and sea. By removing uncertainty over direction, it made longer journeys possible, with particular benefits to the business of import and export.

Something to Think About ...

Muslims, whose faith requires them to pray facing Mecca wherever they are, carry a specially adapted *qibla* compass to indicate the direction of the city.

AD c70

The Chinese apply the directional properties of magnetic lodestones (which they use to plan buildings) to the world at large. A pivoting lodestone is the first compass. By the 11th century a magnetized needle floating in a bowl of water is widely used by Chinese seafarers. Sailors in the West are using a similar device within 150 years.

1908

American entrepreneur Elmer Sperry patents the gyrocompass, a gyroscopic device that indicates true north, not magnetic north, and does not suffer from magnetic interference. It is widely adopted, particularly by large ships, and remains the standard navigational aid until the civilian introduction of the Global Positioning System in the mid-1990s.

C1300

By 1300 Western mariners are using a dry compass – a needle balanced on a pin, sealed in a box behind glass. Later a gimbal, a pivoted support that allows the rotation of an object about a single axis, is added to keep the compass level on rough seas. In 1690 English astronomer Sir Edmund Halley devises a liquid compass whose needle moves in oil or alcohol, not air: this acts as a shock absorber, making it easier to read.

1854

Marine engineer John Gray from Liverpool incorporates adjusting magnets in the binnacle housing of a ship's compass to compensate for interference from the hulls of the new iron ships. A revised version is adopted by the US Navy in 1860, but not by the Royal Navy until 1908. Correction is also necessary when the first liquid compass is installed on a plane in 1909.

Paper_AD 105

Ts'ai Lun (China)

The word 'paper' is derived from papyrus, a plant found in Egypt along the lower Nile from which Egyptians created sheets on which to inscribe. Some 3,000 years later a Chinese man named Ts'ai Lun soaked bamboo fibres and the inner bark of a mulberry tree in water, creating a mixture that, when dried on a flat piece of cloth, became a writing surface. This papermaking process was used in China before being passed along to the rest of the world.

AD 105 Ts'ai Lun from China combines soaked bamboo fibres and the bark of a mulberry tree to produce the first paper.

Something to Think About . . .

The sizing system for paper in Europe is based on common width-to-height ratios for different paper sizes. The largest standard size paper is A0 (A zero), and two sheets of A1 paper placed upright side by side, fit exactly into one sheet of A0 laid on its side.

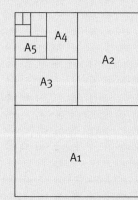

Similarly, two sheets of A2 fit into one sheet of A1 and so on. Common sizes used in the office and the home are A4 and A3, the latter being the size of two A4 sheets.

AD C900 In Arabia linen fibres are substituted for wood and bamboo, creating a higher quality surface.

C1100 Papermaking reaches Europe, where water-powered paper mills are built and the mechanization of papermaking begins.

1448 The invention of the printing press leads to a huge increase in the demand for paper.

1843 Wood replaces textiles as the most common source of fibre for papermaking. At the same time steam-driven machines begin to revolutionize the papermaking process.

Chapter 02.0

Gunpowder to Seed Drill

c900 — 1701

Gunpowder_AD c900

Unknown Alchemist (China)

Strange that an invention so associated with death has so many connections with healing and prolonging life. Strange too, that the inventions of dynamite (1867) and gelignite (1875) fund the Nobel Peace Prize. Gelignite, based on nitroglycerine, was the first plastic explosive. Semtex, a more recent plastic developed in the 1950s, is based on PETN (pentaerythritol tetranitrate). Despite its birth in the Chinese quest for immortality, it is the destructive power of gunpowder and its successors which has left its mark on history.

Gunpowder – a careful mix of potassium nitrate, charcoal and sulphur – is invented by Chinese alchemists seeking eternal life. Instead they produce fireworks ('Chinese flowers') to ward off evil, and firearms ('Chinese arrows') to repel enemies.

Ascanio Sobrero of Turin invents nitroglycerine, to treat cancer and heart conditions. As an explosive it is highly unstable until, in 1867, Alfred Nobel combines it with inert minerals to make dynamite. Nobel's brother Emil is killed during early tests.

Julius Wilbrand develops TNT (trinitrotoluene) as a yellow dye. It proves to be so stable and hard to detonate that it isn't used as an explosive until 1902. During the First World War, yellow staining to the hands of the mainly female munitions workers earned them the nickname 'canary girls'.

Something to Think About . . .

Nitrocellulose, developed from an early explosive, was briefly used as a coating on billiard balls. However, in a game that involves bouncing hard balls off one another with a sharp stick, it was a considerable disadvantage that the covering ball sometimes exploded on impact.

1891

1964

At the University of Chicago, Philip Eaton synthesizes cubane, a highly reactive molecule, previously considered only a theoretical possibility. By 1999 its relatives, hepta- and octa-nitracubane are outperforming standard military explosives by 25 per cent.

PETN (pentaerythritol tetranitrate) is first synthesized by German chemists Tollens and Wigand. Like nitroglycerine it is a treatment for angina, and is the base for many plastic explosives. Its low vapour emission makes it hard to detect.

Spinning Wheel_c1000

Unknown Spinner (Asia)

The story of spinning is a story of a few giant leaps forward. But each new step has built on the pioneering innovations of earlier inventors. Hargreaves' spinning jenny drew heavily on the work of Thomas Highs, and would have been redundant had John Kay not devised the flying shuttle for weavers 30 years earlier, thereby increasing the demand for spun cotton.

The wheel too exists in many forms around the world, each with its own technological advantages. Although it remains a symbol of a pre-industrial age, don't forget that its medieval arrival boosted spinner productivity by a factor of ten, an earlier industrial revolution.

Something to Think About . . .

Some historians argue that the invention of the spinning wheel had a dual impact on civilization. More thread, they argue, meant more cloth, which meant more rags – rags being an important ingredient of mass-produced paper, essential for the spread of printing.

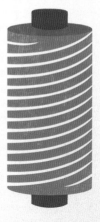

20,000 BC

The earliest evidence of yarn is a crude skirt of rough fibres, probably rolled by hand on the spinner's thigh. The invention of the spindle speeds the process. Its spin is regulated by a weight or whorl, which makes longer, more even threads possible.

1533

The hand-turned spinning wheel may have evolved separately in Asia and China. It spreads to Europe by the 14th century. The addition of a treadle to power the wheel using the foot, probably in 1533 in northern Germany, leaves both hands free for spinning.

1764

In England, Blackburn-based weaver James Hargreaves devises the first machine that is able successfully to turn many spindles at once, known as a spinning jenny or engine. With increased output, the price of yarn collapses, and spinners in Blackburn break into Hargreaves' workshop and smash his machines.

1963

The Czech Cotton Research Institute devises the Open End Spinning process, which in effect spins not the spindle but the cotton, to put the twist in the yarn being pulled from it. It produces relatively coarse threads, at speeds of over 40m (131ft) per second.

Gun_c1250

Unknown (China)

The need for power and speed fired the imagination of the inventors of firearms technology. More efficient explosions threw bullets further. The desire to reload quicker and less often prompted multiple barrels and bullet-chamber revolvers in the 1590s, machine guns from 1851 and lever-action repeaters in 1855. Modern Metal Storm electronic weapons can fire over a million rounds a minute by gas ignition.

Many inventors of new weapons are remembered in the names of their creations – Colt, Gatling, Browning, Smith & Wesson, and Kalashnikov to name a few. Their inventions are associated with the conflicts in which they were used, from the taming of the American West (the Colt 45) to the guerrilla conflicts of the Middle East (Kalashnikov's AK-47).

But not all guns are designed to kill. Lives are saved, for example, by the Verey flare gun; and guns firing nails instead of bullets are commonly used on construction sites to make homes, not war.

Something to Think About ...

Phrases from the early days of firearms have become part of our language. A flash in the pan was one that did not ignite the main gunpowder charge. Flintlocks were designed not to fire if the hammer was only half-cocked. Both terms are commonly used now for an event or person who is ineffective or does not deliver fully on their potential.

Chinese handgun

French flintlock

Winchester repeater

Borchardt automatic

Some key bullet points

In the 13th century the first handguns are fired by lighting a fuse that ignites a small charge of gunpowder. The flash from that lights the main charge in the barrel.

Refinements in the way the gunpowder is lit culminate by 1610 in the flintlock, a simple, reliable mechanism, which strikes a piece of flint against a steel plate to make a spark.

In 1836 the bullet, charge and detonator are all packed together for the first time, in a single cartridge, against which the trigger propels a small hammer or pin to fire it.

The automatic pistol appears in 1893. It harnesses discharge gases or recoil energy to expel the spent round and reload from a magazine built into the grip of the gun.

Spectacles_ 1284

Salvino D'Armate (Italy)

Spectacles, or glasses as they are now most often known, are the most common way to correct short or long sight. Myopia or short-sightedness that frequently occurs in middle age means that light focuses in front of the retina at the back of the eyeball. Long-sightedness, also known as far-sight or hyperopia, means the eyes focus behind the retina, which makes it difficult to see objects close up rather than at a distance.

The objective is to move the focal point on the retina to correspond with the eye's condition, creating apparently perfect vision. This is achieved by varying the curve, the thickness and the shape of the lens – short-sightedness is corrected by concave and long-sightedness by convex lenses.

Something to Think About . . .

They may be fashion accessories these days, but sunglasses are intended to prevent eye damage from the sun's ultraviolet rays. Tinted lenses were first used in the mid-18th century by James Ayscough, but it took until 1929 for Edwin Land to invent a cellophane-like polarizing filter that, in 1937, became the basis of Polaroid sunglasses.

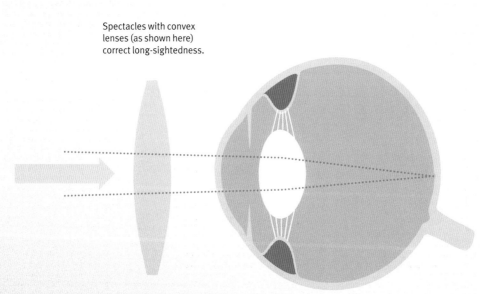

Spectacles with convex lenses (as shown here) correct long-sightedness.

Convex spectacle lenses allow images to form on (not behind) the retina of the long-sighted eye. Concave lenses allow images to form on (not in front of) the retina of the short-sighted eye.

C1000
First instance in Roman records of the reading stone: a glass sphere laid on top of books to magnify the letters.

1284
Italian inventor Salvino D'Armate is credited with inventing the first wearable eyeglasses.

1752
London-based eyeglass designer James Ayscough introduces spectacles as we know them today with hinged side arms.

1784
American Founding Father Benjamin Franklin develops bifocal glasses that allow the wearer to see both up close and at a distance: the distance lens is placed in the upper half of the glasses and the close-work lens in the lower half.

1888
German Adolf Fick invents the first contact lens with refractive power – the ability to bend rays of light. Flick's 'contact spectacle' is a thin glass bowl placed on the eyeball, the area in-between being filled with a saline liquid similar to tears.

Printing Press_1440

Johannes Gutenberg (Germany)

When Johannes Gutenberg made the first printing press with movable type in 1440 he put printed material, formerly an expensive luxury, within the grasp of the masses for the first time. The type – made of metal soft enough to cast yet hard enough to use for printing – was individually cast in moulds and could be taken apart for re-use, while the press itself could print a then unprecedented three sheets per minute. A welcome by-product of this innovation was a trend towards the standardization of spelling.

1609
Germany becomes the home of the world's first weekly newspaper, *Relation*. All early newspapers consist of a single sheet on which text is printed in two columns.

1440
German goldsmith Johannes Gutenberg develops a printing press, using the wooden screw-type wine presses of the Rhine Valley as the basis for his invention. An operator works a lever to increase and decrease the pressure of the metal type against the paper.

1731
Published in England, *The Gentleman's Magazine* is the first periodical to use the word 'magazine' as a term meaning storehouse of knowledge. With stories, poems and political comment, it provides the first regular employment for author Samuel Johnson and continues to be published up until 1922.

Something to Think About . . .

The inventor of the printing press gave his name to the Gutenberg Bible of 1454. This was the first major book printed using movable type, and contained the maximum number of lines per page – at that time 42. The Gutenberg Bible was typical of the religious material that formed the majority of early printed matter.

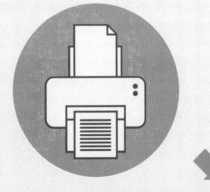

1843

American inventor Richard M. Hoe devises the rotary printing press, marrying steam power to a cylinder on which the type is mounted. This allows millions of pages to be printed daily.

1998

The first eBook readers appear on the commercial market, but would take a decade to catch on. In 2011 Internet retailer Amazon reports sales of digital books for its Kindle reader to be outselling hardbacks by a factor of 1.4 to 1.

1985

Desktop publishing becomes an everyday phenomenon, thanks to the availability of relatively inexpensive laser printers and computers. The Apple Macintosh is the early front-runner.

Pencil_1564

Unknown (England)

The so-called 'lead' pencil was invented in 1564 when a deposit of black carbon was discovered near Borrowdale in Cumbria, England. This improved on the work of the Faber family of Nuremberg, Germany, who had used pulverized graphite to create a prototype.

In 1795, French chemist Nicolas Conte patented a method of kiln-firing powdered graphite with clay to make graphite rods for pencils. The hardness of the graphite could be varied by changing the ratio of graphite to clay – important to artists and draughtsmen.

Something to Think About . . .

The word 'pencil' comes from the Latin word 'penicillus', which means 'little tail', the name of the small brush ancient Romans used as a writing instrument. 'Graphite' was named in 1789 by A. G. Werner after the Greek word meaning 'to write'.

Fountain Pen_1884

American Lewis Waterman patented the first practical fountain pen, with its own built-in ink reservoir, in 1884. His was one of four companies, including Parker and Sheaffer, to dominate the market for the next six decades.

Prior to this, the quill was the principal writing instrument, the best made from goose or swan feathers. These went into decline after the invention of the dip or nib pen, which was patented in America in 1810, but like the quill, had to be dipped into ink.

Romanian Petrache Poenaru received a French patent for inventing the first pen with a replaceable ink cartridge in 1827, but the system took nearly 150 years to catch on.

Ballpoint Pen_1938

Hungarian journalist László Biró invented the first ballpoint pen in 1938 using quick-drying ink dispensed by a tiny ball bearing in its tip. This sphere can range from 0.5–1.2mm in diameter and is usually made of brass, steel or tungsten carbide.

The Royal Air Force produced ballpoints under licence during the Second World War, finding them more effective than leaky fountain pens at altitude. The Germans were denied the technology as Biró fled to Argentina. (See page 164 for more information about the now ubiquitous ballpoint.)

Thermometer_ 1593

Galileo Galilei (Italy)

A traditional thermometer measures temperature by calibrating the change in the mass of a substance, for instance mercury, which expands when it is heated and contracts when it is cooled. The scale typically used is Celsius, with 100 degrees Celsius (°C) between freezing point (0°C) and boiling point (100°C) at sea-level air pressure, invented by Swedish astronomer Anders Celsius (1701–44) and adopted internationally in 1948.

1665
Dutch mathematician Christiaan Huygens suggests using the melting and boiling points of water as standard points to establish a universal temperature scale.

1611
Italian inventor Santorio Santorio is the first inventor to put a numerical scale on the instrument. The thermometer receives its name in 1624 from the Greek words 'thermo' (warm) and 'meter' (measure).

1593
Italian Galileo Galilei invents a water thermometer that allows temperature variations to be measured for the first time. The technical term for an uncalibrated instrument like this is a thermoscope.

1643
Evangelista Torricelli, an Italian physicist and mathematician, invents the barometer, a related instrument used to measure atmospheric pressure.

Measuring our progress

1724
Fahrenheit introduces the temperature scale that bears his name, made possible due to the accuracy of his mercury thermometer. It ranges from 32°F (freezing) to 212°F (boiling).

1970
The first digital thermometer is patented in Alabama by the Royal Medical Corporation.

1714
Gabriel Fahrenheit, a physicist born to German parents in Danzig (now northern Poland) who lived most of his life in the Netherlands, invents the first mercury thermometer, choosing the metal because it has a high coefficient of expansion.

Something to Think About...

Thermometers are used to monitor temperatures in everything from fish tanks to nuclear reactors. For work with high temperatures it may only be possible to measure to the nearest 10°C, while clinical thermometers and many electronic thermometers are readable to 0.1°C.

Flush Toilet_1595

Sir John Harington (England)

The invention of the flush toilet (or water closet as it was known until the 19th century, hence the commonly used abbreviation 'WC'), using mains water to dispose of human waste, has probably done as much to ensure the health of the world as any vaccine. It relies upon a system of sewers to remove the water to treatment plants where the waste can be broken down – areas where this is not practical may employ septic tanks to store waste before disposal.

Something to Think About ...

American inventor Joseph Gayetty's Medicated Paper was the first packaged toilet paper available from 1857. In 1880, the British Perforated Paper Company offered boxes of small pre-cut squares. In 1879, the Scott Paper Company began selling toilet paper on a roll, but this did not catch on until 1907.

800 BC
King Minos of Crete is the owner of the first recorded flushing water closet in his palace at Knossos. A jug is used to flush the drain.

1595
English author Sir John Harington installs an early working prototype of the flush toilet in Richmond Palace for his godmother Queen Elizabeth I. He publishes a pamphlet called *The Metamorphosis of Ajax* with instructions on how to build one.

1829

The Tremont Hotel in Boston, Massachusetts, becomes the first hotel to feature indoor plumbing with eight water closets. Until 1840, indoor plumbing is found only in hotels and the homes of the rich.

1778

British inventor and locksmith Joseph Bramah substitutes the slide valve with a hinged flap that seals the bottom of the bowl.

1880s

Thomas Crapper, a plumber from London, is awarded nine patents for plumbing innovations, three of which are improvements to the flushing water closet.

1775

The first patent for the flushing toilet is issued to Scottish watchmaker Alexander Cummings. He invents the 'S-trap' using the water in the bowl as a seal against foul smells. His design also features a sliding valve in the bowl outlet above the trap.

1885

British potter Thomas Twyford builds the first trapless toilet in a one-piece china design. From now on, china replaces the more common materials of metal and wood in toilet construction.

02.9 Microscope_1608

Hans Lippershey (The Netherlands)

The traditional 'light' microscope, which came about when multiple lenses were used to magnify tiny objects, was superseded in the 1930s by the electron microscope. There are two types, the Transmission Electron Microscope (TEM) and the Scanning Electron Microscope (SEM), and both are able to magnify objects many times greater than the light microscope by employing a focused beam of electrons instead of light. Resolution in the light microscope is limited by the wavelength of the light used, about 250 nanometres (nm) or millionths of a millimetre. As the wavelength of the electron is far shorter, resolutions of 0.3nm are possible at magnifications that go to 1,000,000x.

Something to Think About . . .

Though Galileo was not the inventor of the microscope, he improved it greatly by adding a focusing device. His friend, Johannes Faber, gave the name 'microscope' to Galileo's instrument from the Greek words for 'small' and 'see'; Galileo had called it the 'occhiolino' or 'little eye'.

1608 Dutch spectacle-maker Hans Lippershey attempts to patent the two-lens optical refracting telescope, intending it for military use. The maximum magnification obtained was only about 10x.

1624 In Italy Galileo uses the same two-lens principle to make a compound microscope.

1668 Anthony van Leeuwenhoek from The Netherlands, the 'Father of Microscopy', learns how to make small but highly curved lenses by grinding and polishing. These rounder lenses produce greater magnification and can magnify up to 270x. It is now possible to see things never before observed with the naked eye, including bacteria, sperm, blood cells and a selection of protozoa.

1847 Carl Zeiss starts producing microscopes in Jena, Germany, and his company becomes known for producing fine optical instruments.

1935 German engineer Max Knoll develops the Transmission Electron Microscope. This projects a beam of electrons on to a thin specimen coated with an electrically conductive material, commonly gold. The image returns as a map of the intensity of the signal being emitted from the scanned area.

Pendulum Clock_1656

Christiaan Huygens (The Netherlands)

The concept behind the pendulum clock is that the weight on the end of a rod swings back and forth at a specific and precise time interval dependent on the length of that rod. The pendulum is given the impulses necessary to keep it swinging by an escape wheel or escapement, which gives the characteristic 'tick-tock' sound. The rotation of the escapement is reflected by hands telling the time on a clock face. Any movement would affect pendulum motion, so this mechanism was unsuitable for portable timepieces. Pendulum clocks remained the world standard for accurate timekeeping for nearly three centuries until the invention of the quartz clock in 1927.

Something to Think About . . .

Big Ben, the clock tower at London's Palace of Westminster, is the most famous pendulum clock in the world. Its pendulum, installed within an enclosed windproof box beneath the clock room, is 3.9m (12 ¾ ft) long, weighs 300kg (661lb) and swings every two seconds.

Italian astronomer Galileo Galilei studies pendulum motion, but dies before his design for a clock can be built.

American engineer Warren Morrison develops the first quartz clock at the Bell Telephone Laboratories, setting new standards in accuracy.

1582

1927

1656

1671

1675

Dutch scientist Christiaan Huygens makes the first pendulum clock. Described in his 1658 article *Horologium*, it is regulated by a mechanism with a natural period of oscillation and has an error of less than one minute a day.

Huygens develops a spring assembly for watches, which substitutes a wound-up spring regulated by a balance wheel, for pendulum motion. He presents the first one to be produced to King Louis XIV of France.

A new type of clock is developed by Briton William Clement giving even greater accuracy within a few seconds per day. It includes an anchor escapement that swings back and forth with the pendulum, and 'pallets' at each end of the anchor, which alternately catch and release the escape wheel.

Steam Engine_ 1698

Thomas Savery (England)

Ancient Greek mathematician and engineer Hero of Alexandria is credited with the earliest use of steam power. In the 1st century AD he used two escaping jets of steam to spin a sealed and heated vessel, the first boiler. Not until the late 17th century did inventors seek to harness the properties of steam for practical purposes, and since then only the development of the electric motor and the internal combustion engine have challenged its supremacy.

1698
English military engineer Thomas Savery devises a crude steam-driven pump for draining flooded mines. Unfortunately the boiler tends to blow up under pressure.

Something to Think About ...

Vive la Revolution! The Industrial Revolution, during which steam engines changed the world, owes its success to a French pressure cooker. French physicist Denis Papin devised his 'steam digester' in 1679 to separate fat from bones. The steam-release valve, which he introduced after a series of explosive experimental models, directly inspired Thomas Savery to develop that first steam-powered water pump.

1712
English blacksmith Thomas Newcomen builds a piston-driven pump – hot steam pushes the piston one way, and when the steam is cooled the resulting vacuum pulls the piston back. Unfortunately the cooling and reheating required make it very inefficient.

1769

Scottish mechanic James Watt adds a separate condenser to solve Newcomen's problems. The energy conserved by being able to keep the engine hot and the condenser cool saves about 75 per cent in fuel consumption. This time there is no 'unfortunately' – steam engines based on Watt's version power the Industrial Revolution around the world.

1884

English nautical engineer Charles Parsons patents a steam turbine, driving a rotor instead of a piston, and applies it to shipping and the generation of electricity.

1800

Cornish mining engineer Richard Trevithick perfects a high-pressure engine. There is no need for a condenser, and it can deliver the same power with a smaller piston cylinder. As a result of this early example of miniaturization, steam railways become possible.

Seed Drill_1701

Jethro Tull (England)

Before seed drills, seed was scattered in the fields by hand. Although the fields were prepared by ploughing, much of the seed fell outside the furrows. There it would be exposed to scavenging wildlife, and even if it took root, its roots were likely to be shallower and its growth stunted. Seed drills maximize the yield from the field through efficient spacing and planting. The uniformity of the rows makes them easier to keep clear of weeds, so that more nutrition is left in the soil for the crop itself.

Jethro Tull applied his inventive mind to many aspects of crop farming. He argued for the use of horses instead of oxen in the fields as they were more maneuverable and could operate between rows more effectively. In addition to the seed drill, he invented a new horse-drawn hoe; and he made significant improvements in the design of the traditional plough. Although his workers felt that their jobs were threatened by these new machines, they were adopted by landowners over time, and their use transformed the English countryside forever.

Something to Think About . . .

The rock band Jethro Tull was known by many different names in its early days as they had trouble getting repeat bookings and they found that changing the band's name enabled them to remain on the London club circuit. They were given the name 'Jethro Tull' by a member of their booking agency who happened to be a history enthusiast, and this was the name they were using when they received their first repeat booking.

1701

English agriculturalist Jethro Tull studies European farming methods and designs a horse-drawn seed drill. His scientific approach to farming starts the Agricultural Revolution.

1566

Three hundred years after Marco Polo's travels to the east, Venetian Camillo Torello invents an early European version of the seed drill, possibly based on reports of the Chinese model.

c200 BC

The Chinese use a formative ox-drawn version of the seed drill, capable of planting several rows at once. This forerunner of the seed drill is credited with their success at feeding the country's huge population.

c1500 BC

The Sumerians, some of the earliest crop farmers, develop a simple, single-row seed planter. This invention does not spread beyond their borders in southern Mesopotamia (Iraq).

63

Chapter 03.0

Screw-cutting Lathe to Telegraph
1775 — 1836

03.1 Screw-cutting Lathe_1775

Jesse Ramsden (England)

The humble nut and bolt are often overlooked, but the success of the Industrial Revolution hinged on their evenly spaced threads. Without their standardized pitches, angles and diameters, different factories would not have been able to work together on separate components of the same product.

With interchangeability came compatibility, speed of assembly and mass production. The screw-cutting lathe ranks with the steam engine as one of the crucial inventions that made the Industrial Revolution possible. It could also be blamed for the size of today's DIY industry!

Something to Think About . . .

The earliest screws were wooden components used to squeeze the grapes in wine presses, the same technology that was later applied to Gutenberg's first printing press.

From da Vinci to DIY

Leonardo da Vinci (1500)

Leave it to da Vinci! Around 1500, he draws plans for a screw-cutting lathe (probably never built), adding a flywheel to give it uniform speed and direction, and two threaded rods or lead screws to guide the tool evenly along the work.

The lathe (1300 BC)

The ancient Egyptians use lathes as early as 1300 BC. These lathes are hard to adapt to screw-cutting because they switch direction – clockwise then anti-clockwise. It is difficult to cut a consistent thread.

The screw-cutting lathe (1775)

In 1775, Jesse Ramsden, pioneering scientific instrument maker, makes a new screw-cutting lathe, the first ever to use the lead screws devised by da Vinci. Refinements by Henry Maudslay in 1800 make possible the standardization of screws.

The screw (100 BC)

The first use of a screw as a way of holding things together is around 100 BC. It is more secure than a nail, but hard to make by hand, with no guarantee that the nut made for one bolt will fit the thread of another.

The assembly line (1913)

Interchangeable nuts and bolts are now used to connect interchangeable parts in products that have previously been individually handcrafted. This dramatically reduces assembly time and labour costs, as Henry Ford realizes in 1913.

Hot-Air Balloon_1783

Joseph and Jacques-Étienne Montgolfier (France)

The first free flight by men was in a taffeta balloon filled with air that was heated by burning wool, designed by the Montgolfier brothers. Joseph was inspired by watching laundry billowing as it dried by the fire.

Étienne had been the first to leave the ground, in a tethered balloon at their workshop in Paris a month earlier, and a month before that a sheep, a duck and a cockerel were the first experimental passengers in a balloon flight.

21 November 1783

The Marquis François d'Arlandes and Jean-François Pilâtre de Rozier are the first humans to fly by hot-air balloon across Paris, covering 9km (5 ½ miles) in 25 minutes.

1 December 1783

Ten days later, the first gas-filled balloon flight by Jacques Charles lasts 150 minutes and carries him 43km (26 ¾ miles) across Paris, watched by half the city's population.

Something to Think About . . .

The earliest known use of unmanned balloons was by the Chinese in the 3rd century AD. They floated paper lanterns into the air as military signals.

24 September 1852

The first steerable airship flies for 27km (16 ¾ miles) when French engineer Henri Giffard attaches a small steam engine to a propeller, guiding his balloon from Paris to Trappes.

26 November 2005

Indian businessman and aviator Vijaypat Singhania sets the world altitude record for a hot-air balloon flight. From central Mumbai, India, he rises to 21,027m (68,969ft) and lands in Panchale 240km (149 miles) to the south.

20 March 1999

Balloonists Betrand Piccard and Brian Jones are the first to circumnavigate the globe non-stop, taking 19 days, 21 hours and 47 minutes in the balloon *Breitling Orbiter 3*.

16 August 1960

Highest parachute jump from any aircraft, when US Air Force pilot Joe Kittinger leaps from *Excelsior III*, a helium balloon 31,333m (102,800ft) above New Mexico.

2 July 1900

First untethered flight of a rigid-framed airship, designed by Graf Zeppelin, a German general and later an aircraft manufacturer, near Lake Constance in Germany. Future Zeppelins will cross the Atlantic Ocean.

Electrical Cell_ 1800

Alessandro Volta (Italy)

Alessandro Volta's wet-cell battery was called a voltaic pile. It consisted of discs of zinc and copper separated by pieces of cardboard soaked in brine. Electrical current was produced by the contact of different metals, brass and iron, in a moist environment.

The use of batteries has developed exponentially over the years, with the increases in useful life and variety of sizes on offer being the two significant factors. They now power everything from hearing aids to portable MP3 players, watches to mobile phones, and are a mainstay of our modern world.

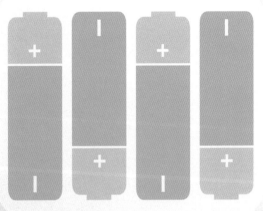

Something to Think About . . .

The term 'battery' was first used in 1748 to describe an array of charged glass plates. Volta gave his name to the volt, the unit of electromotive force or difference of potential that causes a current of one ampere to flow through a resistance of one ohm.

1800

Alessandro Volta invents the first wet-cell battery to produce a steady current of electricity.

1836

Englishman John F. Daniell invents a cell that uses two electrolytes, copper sulphate and zinc sulphate. The Daniell cell is somewhat safer and less corrosive than the Volta cell.

1839

William Robert Grove develops the first fuel cell, which produces electricity by combining hydrogen and oxygen.

1859

French inventor Gaston Planté develops the secondary cell, a lead-acid battery that can be recharged. This type of battery is still used in cars today, and the secondary cell is the basis of today's rechargeable battery.

1866

In France, engineer Georges Leclanché patents the carbon-zinc wet-cell battery: a zinc anode and a manganese dioxide cathode wrapped in a porous material and submerged in a jar of ammonium-chloride solution. These power early telephones.

1881

German Carl Gassner invents the first commercially successful dry-cell battery. The zinc-carbon battery is still around today, in spite of its relatively short life.

1901

In America, inventor and scientist Thomas Alva Edison devises the alkaline storage battery, originally too expensive for commercial use but now the most common battery type.

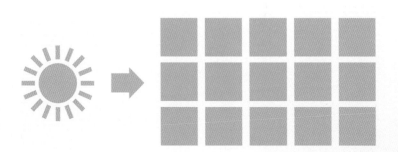

Solar Battery_ 1954

American researchers at Bell Labs, Gerald Pearson, Calvin Fuller and Daryl Chapin, invent the first solar battery by placing an array of silicon strips in sunlight. These capture free electrons and turn them into electrical current.

Canned Food_1810

Auguste de Heine and Peter Durand (England)

At the beginning of the 19th century Napoleon Bonaparte saw that traditional methods of preserving food did not keep it edible for long enough to reach France's far-flung armies and offered a reward to anyone who could come up with a new method. Soon after this, it was discovered that iron canisters could be used, and these had the advantage over glass bottles of being lighter, easier to seal and more durable during transportation. The iron was coated with a fine layer of tin to stop it from rusting, hence the name 'tin can'.

Something to Think About ...

The first cans were sealed with lead solder, and this put the people who ate food from them in danger of developing lead poisoning. Every member of the Arctic expedition led by Sir John Franklin in 1845 died from severe lead poisoning after eating from cans like this for three years.

1808

French confectioner and chef Nicolas Appert discovers that heating food to high temperatures inside sealed glass jars stops it from 'going off'.

1810

Two inventors, Auguste de Heine and Peter Durand, take out patents on iron and tin containers for preserving food. These 'cans' weigh more than the food they hold.

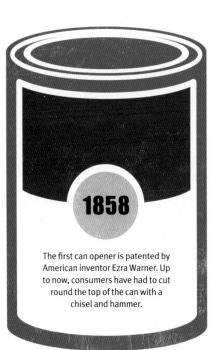

1858

The first can opener is patented by American inventor Ezra Warner. Up to now, consumers have had to cut round the top of the can with a chisel and hammer.

1901

The American Can Company is founded in Greenwich, Connecticut. In the early part of the 20th century they produce 90 per cent of the tin cans used in the USA.

1962

The ring pull, which enables tin cans to be opened without a can opener, is invented by American Ermal Fraze. It is licensed to the Pittsburgh Brewing Company, which use it on beer cans.

Frozen Food_1929

American inventor Clarence Birdseye develops the quick-freezing system, packing fresh food into waxed cardboard boxes and flash-freezing it under high pressure with an electric fan, buckets of salt water and cakes of ice. The first quick-frozen vegetables are presented to the public in 1929.

Bicycle_1817

Karl von Drais (Germany)

Little has changed in bicycle design since John Kemp Starley gave it its modern form in 1885. Technology and new materials have been adapted to give advantages of speed and comfort, and variations such as the BMX bike have evolved to cater for specific ways of using the bicycle for fun. Recumbent bicycles first appeared in the 1890s, but their development was restricted in 1934 when they were banned from competitions for being too fast! In 2009 a recumbent cyclist set the human-powered land speed record of 134km/h (83mph).

Hobbyhorse (1817)

In 1817 Baron von Drais invents a two-wheeled device for getting around his estates. Its frame is wooden, and he can steer it by the front wheel, but it has no pedals – his feet touch the ground and he balances astride the machine while walking or running.

Boneshaker (1860s)

During a Parisian craze for bicycles in the 1860s, blacksmith Ernest Michaux is one of several claiming to have added pedals for the first time. They are on the front wheel, which restricts steering. The nickname comes from riding the bike on cobbled streets.

Something to Think About...

John Kemp Starley's first modern bicycle was named the Rover, and so was the company he formed to manufacture it. After his death the company began to produce motorcycles and eventually motor cars – the Rover cars of which today's Land Rover and Range Rover models are the modern descendents.

Penny farthing (1870–71)

Around 1870 the boneshaker's ride is smoothed out and speeded up by increasing the diameter of the front wheel to as much as 1.5m (5ft). It's a long way down for a falling rider! Coventry engineer James Starley adds tangential spokes in 1871.

Safety bicycle (1885)

Henry Lawson first tries a rear-wheel chain drive on a penny farthing in 1879, but it is James Starley's nephew John Kemp Starley who produces the first modern bicycle with wheels of equal size, handlebars, pedals, chain and a sprung saddle, in 1885.

Braille_ 1824

Louis Braille (France)

Braille is the unique raised writing that enables visually impaired people to 'read'. It was developed by Frenchman Louis Braille who became blind as a result of an accident at the age of three.

Each character is based on a cell with six dots in a rectangular shape, two across and three deep. The number of dots and their position determine the letter, number or symbol of the character. The 64 possible permutations allow Braille to represent a variety of languages. Computer technology may replace Braille in our lifetime, but two centuries after its invention it is still the primary form of reading and writing for the visually impaired.

Something to Think About ...

Braille is commonly used for safety reasons on cardboard and plastic medicine packaging, while the concept of raised markings has also been used by Mexico's central bank to make notes distinguishable from one another.

A	B	C	D
E	F	G	H
I	J	K	L
M	N	O	P
Q	R	S	T
U	V	W	X
Y	Z		

Early 1800s

French army captain Charles Barbier de la Serre creates a forerunner of Braille called night writing. Intended to enable soldiers to communicate noiselessly and in poor light, night writing is considered too complex for them to learn.

1821

Barbier visits the National Institute for the Blind and meets 12 year old Louis Braille. By 1824, aged 15, Braille has modified night writing for use as a communication tool for blind people.

1919

The Universal Braille Press (now the Braille Institute) is founded. This currently produces more than five million Braille pages annually.

1945

The National Braille Association (NBA) is set up to provide education and Braille materials to persons who are visually impaired.

1976

American inventor Raymond Kurzweil perfects the Kurzweil Reading Machine, which uses a flatbed scanner and optical character recognition software to feed a speech synthesizer. He goes on to develop a speech-recognition system.

LOUIS BRAILLE FRANCE

BRAILLE IS THE UNIQUE RAISED WRITING THAT ENABLES VISUALLY IMPAIRED PEOPLE TO READ. IT WAS DEVELOPED BY FRENCHMAN LOUIS BRAILLE WHO BECAME BLIND AS A RESULT OF AN ACCIDENT AT THE AGE OF THREE.

EACH CHARACTER IS BASED ON A CELL WITH SIX DOTS IN A RECTANGULAR SHAPE, TWO ACROSS AND THREE DEEP. THE NUMBER OF DOTS AND THEIR POSITION DETERMINE THE LETTER, NUMBER OR SYMBOL OF THE CHARACTER. THE 63 POSSIBLE PERMUTATIONS ALLOW BRAILLE TO REPRESENT A VARIETY OF LANGUAGES. COMPUTER TECHNOLOGY MAY REPLACE BRAILLE IN OUR LIFETIME, BUT TWO CENTURIES AFTER ITS INVENTION IT IS STILL THE PRIMARY FORM OF READING AND WRITING FOR THE VISUALLY IMPAIRED.

SOMETHING TO THINK ABOUT . . .

BRAILLE IS COMMONLY USED FOR SAFETY REASONS ON CARDBOARD AND PLASTIC MEDICINE PACKAGING. WHILE THE CONCEPT OF RAISED MARKINGS HAS ALSO BEEN USED BY MEXICO'S CENTRAL BANK TO MAKE NOTES DISTINGUISHABLE FROM ONE ANOTHER.

EARLY 1800S

FRENCH ARMY CAPTAIN CHARLES BARBIER DE LA SERRE CREATES A COMMUNICATION SYSTEM CALLED NIGHT WRITING. IT WAS INTENDED FOR SOLDIERS TO COMMUNICATE NOISELESSLY AND IN POOR LIGHT, BUT NIGHT WRITING WAS CONSIDERED TOO COMPLEX FOR SOLDIERS TO LEARN.

1821

BARBIER VISITS THE NATIONAL INSTITUTE FOR THE BLIND AND MEETS 12 YEAR OLD LOUIS BRAILLE. BY 1824, AGED 15, BRAILLE HAS MODIFIED NIGHT WRITING FOR USE AS A COMMUNICATION TOOL FOR BLIND PEOPLE.

1957

THE UNIVERSAL BRAILLE PRESS NOW THE BRAILLE INSTITUTE IS FOUNDED. THIS CURRENTLY PRODUCES MORE THAN FIVE MILLION BRAILLE PAGES ANNUALLY.

1975

THE NATIONAL BRAILLE ASSOCIATION NBA IS SET UP TO PROVIDE EDUCATION AND BRAILLE MATERIALS TO PERSONS WHO ARE VISUALLY IMPAIRED.

1980

AMERICAN INVENTOR RAYMOND KURZWEIL PERFECTS THE KURZWEIL READING MACHINE, WHICH USES A FLATBED SCANNER AND OPTICAL CHARACTER RECOGNITION SOFTWARE TO FEED A SPEECH SYNTHESIZER. HE GOES ON TO DEVELOP A SPEECH RECOGNITION SYSTEM.

Match_1826

John Walker (England)

Fire has been central to mankind's survival for nearly two million years, for warmth, for light, to scare away predators and to cook food. Between 400,000 and 100,000 BC, we learned to light fires at will. The ability to make a flame has remained a priority even in the age of electricity, whether we're igniting space rockets, fireworks or dinner candles.

Early modern matches were full of white phosphorus, poisonous to both their users and the factory workers who made them. There was enough white phosphorus in one box of matches to kill someone. The first matchbooks appeared in the 1890s, and the hobby of collecting matchbooks and matchboxes is known as 'phillumeny', from the Greek words for 'love' and 'light'.

c400,000 BC · **Rubbing sticks together** This primitive method of making fire involves the production of intense heat from the friction of turning one stick against another to light a combustible tinder material such as wool or leaves.

c1450 · **Match-cord** The word 'match' comes from 'myxa', the Latin word for a fuse. A smouldering length of rope impregnated in chemicals is kept ready to fire cannons or light fireworks.

1826 · **Friction match** Chemist John Walker produces the first modern match. A later manufacturer, Samuel Jones, calls them lucifers; they are smelly and violent until reformulated in 1836.

1913 · **Cigarette lighter** Using 15th-century flintlock musket technology, a spark from a metal flint ignites a flammable vapour of naphtha (an oil distilled from coal) or butane (a hydrocarbon gas). Ronson were first company to do this in 1913; Zippo in 1932.

Something to Think About . . .

A well-known urban myth is the story of the man who saved the
match company a fortune by suggesting they only put sandpaper
on one side of the matchbox. By that simple act the company
halved its annual sandpaper bill. Now that's what you call
thinking outside the box!

Camera_1827

Joseph Nicéphore Niépce (France)

The camera obscura
In 1827 French inventor Joseph Nicéphore Niépce's heliographs (sun prints), are the prototype for the modern photograph, letting light create the picture. However, Niépce's photograph requires eight hours of exposure to create and soon fades.

Something to Think About . . .
Kodak stopped producing Kodachrome roll film in 2009 after 74 years, digital photography having reduced demand. It had already, however, been immortalized by Paul Simon's 1973 song of the same name praising its 'nice bright colours' and prophetically ending 'Mama, don't take my Kodachrome away'.

1827

The Daguerreotype
Developed by French artist Louis Daguerre in 1837, this is the first permanent image. A sheet of silver-plated copper coated in iodine, creating a light-sensitive surface, is inserted into the camera. It is exposed for a few minutes and then bathed in a silver chloride solution, creating an image that doesn't fade when exposed to light.

1837

The film camera
In 1884 American George Eastman patents a flexible, paper-based photographic film that makes the mass-produced camera a reality. The Kodak 'box' camera, a wooden, lightproof box with a simple lens and shutter, follows in 1888. The photographer pushes a button to produce a negative and this is returned to the company for processing. The film was soon changed to a more flexible plastic base.

1884

35mm cameras
German optical engineer Oskar Barnack devises the idea of reducing the size of film negatives and enlarging the photographs in a darkroom after they have been exposed. The world's first 35mm camera, the 'Ur-Leica', is marketed in 1925, its film strip adapted from Edison movie stock.

1925

Flash photography

Flashlight powder is invented in Germany in 1887 by Adolf Miethe and Johannes Gaedicke. The first flashbulb, invented by Austrian Paul Vierkotter, uses magnesium-coated wire in an evacuated glass globe. The first commercially available flashbulb is patented in 1930.

Colour photography

Kodachrome, introduced in 1935, is the first colour still film brought to the mass market. Uniquely, the dyes are not in the film but in the individual chemical baths (yellow, cyan, magenta) and are chemically bonded to the emulsion during processing. The process is so complex that Kodachrome is initially sold at a price that includes processing.

Instant photography

American inventor and physicist Edwin Land invents the one-step process for developing and printing photos that becomes known as instant photography. Land's Polaroid camera is first sold to the public in 1948.

Digital camera

In 1975 Steven J. Sasson takes a black-and-white image with the world's first digital camera. Captured at a resolution of .01 megapixels, it takes 23 seconds to record on to a digital cassette tape and the same time to read off a playback unit on to a television. The first professional digital system is marketed by Kodak in 1991; consumer products follow three years later.

1930

1935

1948

1975

Public Railways_1829

George Stephenson (England)

George Stephenson is often described as the Father of the Railways. He learned his engineering skills on rail systems in northeastern England that connected coal mines to sea ports. By the time he was appointed as engineer of the Liverpool to Manchester Railway, the world's first intercity passenger line, he was a master of his craft.

The steam engines he designed for that company solved all the technical problems that had plagued earlier efforts, ever since Richard Trevithick's first steam engine of 1800. Stephenson improved the efficiency of the boiler and of the transfer of power from the pistons to the wheels, so his trains ran cheaper and faster. His engineering solutions are still applied to steam engines today.

It was Stephenson's locomotive 'the Rocket' that won him the Liverpool–Manchester job, by demonstrating its efficient running at competitive trials in 1829. Unfortunately, it was also the Rocket (on 15 September 1830, when the intercity line opened to the public) that caused the world's first fatal railway accident. A local politician lost his footing and fell beneath its wheels.

Something to Think About ...

When the Swansea and Mumbles Railway in Wales closed in 1960, it was the longest-running passenger railway in the world, having opened in 1806. It holds another railway world record, for the greatest number of different forms of traction: carriages have been pulled on its tracks by horse, steam, electricity, petrol, diesel and even sail.

Empire State Express No. 999
Country: USA
Date: 1893
Max speed: 131km/h
(82mph)

Stephenson's Rocket
Country: UK
Date: 1830
Max speed: 48km/h
(30mph)

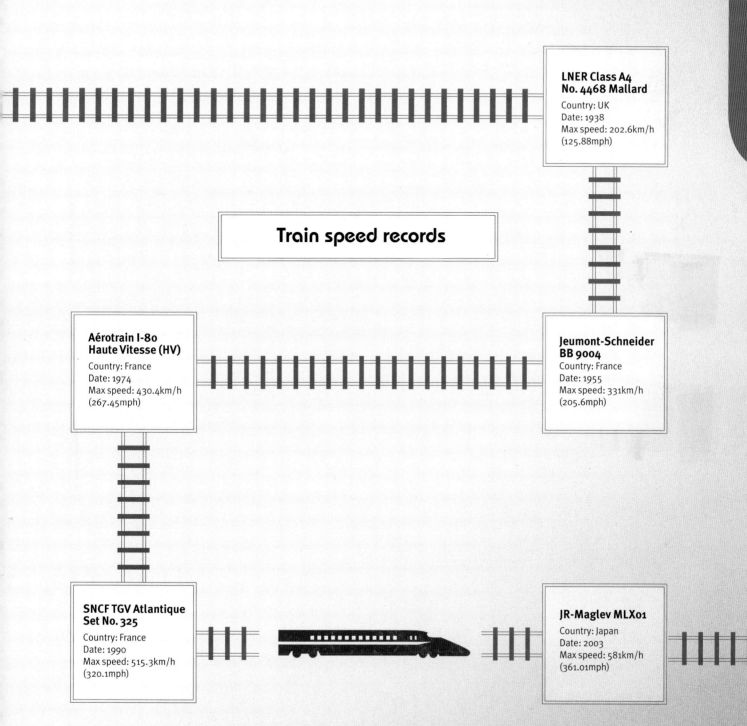

Train speed records

**LNER Class A4
No. 4468 Mallard**
Country: UK
Date: 1938
Max speed: 202.6km/h
(125.88mph)

**Aérotrain I-80
Haute Vitesse (HV)**
Country: France
Date: 1974
Max speed: 430.4km/h
(267.45mph)

**Jeumont-Schneider
BB 9004**
Country: France
Date: 1955
Max speed: 331km/h
(205.6mph)

**SNCF TGV Atlantique
Set No. 325**
Country: France
Date: 1990
Max speed: 515.3km/h
(320.1mph)

JR-Maglev MLX01
Country: Japan
Date: 2003
Max speed: 581km/h
(361.01mph)

Lawnmower_1830

Edwin Beard Budding (England)

The lawn became a popular element of French garden design in the Middle Ages. But it was the spread in the 19th century of public parks and open-air sport – tennis, cricket, croquet and the rest – that created the need for a fast and efficient way of cutting grass.

Budding's invention, and the early mowers that followed, were designed for large recreation areas. Wide-gang mowers comprising several linked cylinders, pulled across golf courses or football pitches by horses or tractors, were the first ride-on mowers. A growing Victorian middle class, who wanted gardens for their homes but could not afford to employ an army of gardeners, began the demand for smaller domestic machines.

Something to Think About . . .

Ride-on mowers took the last bit of physical exercise out of grass cutting – the walking. But since the late 20th century, remote-control machines and robotic mowers mean that armchair gardeners don't even have to be in the garden while their lawn is cut.

Ways of cutting your lawn

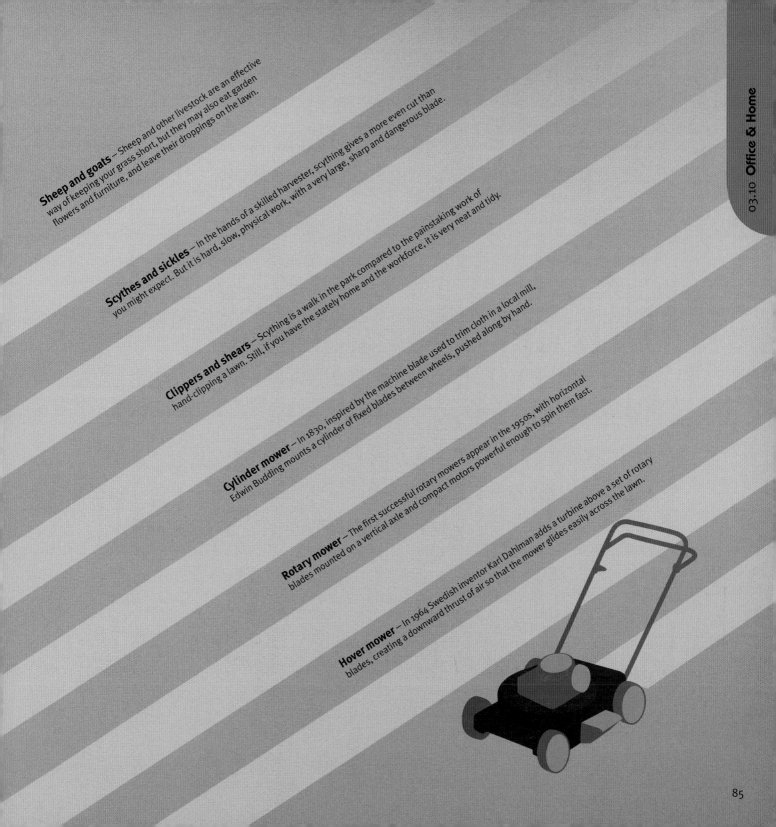

Sheep and goats – Sheep and other livestock are an effective way of keeping your grass short, but they may also eat garden flowers and furniture, and leave their droppings on the lawn.

Scythes and sickles – In the hands of a skilled harvester, scything gives a more even cut than you might expect. But it is hard, slow, physical work, with a very large, sharp and dangerous blade.

Clippers and shears – Scything is a walk in the park compared to the painstaking work of hand-clipping a lawn. Still, if you have the stately home and the workforce, it is very neat and tidy.

Cylinder mower – In 1830, inspired by the machine blade used to trim cloth in a local mill, Edwin Budding mounts a cylinder of fixed blades between wheels, pushed along by hand.

Rotary mower – The first successful rotary mowers appear in the 1950s, with horizontal blades mounted on a vertical axle and compact motors powerful enough to spin them fast.

Hover mower – In 1964 Swedish inventor Karl Dahlman adds a turbine above a set of rotary blades, creating a downward thrust of air so that the mower glides easily across the lawn.

Refrigerator_ 1834

Jacob Perkins (USA)

A refrigerator works by extracting the heat from the fridge or freezer compartment. It does this by compressing liquefied gas in a closed system. The gas vaporizes, drawing energy from its surroundings in the form of heat and therefore lowering the temperature. The gas then passes through a series of coils, which act as a condenser, lowering the temperature of the vapour and making it liquid again, ready to start the cycle once more.

The problem, however, is that gases that liquefy at such low temperatures tend to be dangerous ones. Cullen used ether, then a common general anaesthetic. In the mid-19th century fridges contained ammonia, which is corrosive to the lungs, skin and eyes. By the 1920s sulphur dioxide (which also causes skin burns and loss of vision) and methyl chloride (poisonous and highly flammable) were common refrigerants. Their replacement, low-toxicity Freon, is a chlorofluorocarbon (CFC) now recognized as harmful to the ozone layer. What price our cool beers and frozen food?

Something to Think About . . .

In the rush to solve the problems of refrigeration there were, by 1880, over 3,000 related patents in the USA alone. A late but valuable contribution was made in 1926 by physics genius Albert Einstein. The Einstein Refrigerator had no moving parts, did not need electricity and was driven by a process of heat absorption instead of compression. Its patents were bought by the Electrolux company.

Around 1000 BC the Chinese begin to store ice from frozen lakes in icehouses on the shore; insulated with straw, the ice preserves food and will last until summer.

In 1756 Scottish physician William Cullen gives the first public demonstration of artificial refrigeration by vaporizing ether and making a small ice cube.

Even in the 19th century metal iceboxes are common in many homes. They are kept in a stone room, lined with cork or sawdust and supplied with ice by horse and cart.

In 1834 Anglo-American inventor and mechanical engineer Jacob Perkins patents his 'apparatus for producing ice and cooling fluids', the first working refrigerating machine using vapour compression.

In Australia, Scottish émigré James Harrison develops the first industrial ice-making machine and refrigeration system in 1857, used in brewing and meat packing.

Telegraph_ 1836

Samuel Morse (USA)

Two earlier events made the electric telegraph possible: William Sturgeon's invention of the electromagnet in 1825, and Joseph Henry's development of the electrical relay in 1835. Samuel Morse's was the most successful of several competing systems. When a trial telegraph line was eventually built, its first message, carried from Annapolis to Washington DC in 1844, was the name of the Whig Party presidential candidate. The first transcontinental telegraph, built by Western Union, was completed in 1861. Its rival, the Pony Express, went out of business two days later. The first transatlantic cable was laid in 1866, and suddenly the world seemed much smaller. Western Union finally closed its telegram service in 2006.

Drums

Used in Sri Lanka to communicate between rulers and their subjects since 3000 BC. Widely adopted by armies because the sound can be heard above the noise of battle.

Signal lamps

Light-reflecting mirrors were first used by the ancient Greeks in the 5th century BC. Naval vessels still use powerful lamps to signal at sea when radio silence is necessary.

Can you hear me?

Methods of long-distance communication

Semaphore

By the 18th century many countries had their own systems of visual signals passed quickly from one relay station to the next over long distances by flags or levers.

Something to Think About ...

In 1821 a semaphore signal could be relayed from Paris to Strasbourg via 50 sub-stations in around six minutes – a distance of 362km (225 miles). That's a speed of 3,621km/h (2,250mph), rather faster than the only alternative, a messenger on horseback.

Smoke

Known to have been used by the ancient Chinese to communicate along the Great Wall. Letters found in 1935 describe smoke signals in a Babylonian siege of 588 BC.

Chapter 04.0

Postage Stamp to Microphone
1840 — 1876

Postage Stamp_1840

Sir Rowland Hill (England)

A penny postal system was first introduced in England in 1680, paid for by the recipient of the letter. By 1840 this had risen to as much as fourpence. The Penny Black was just one of Sir Rowland Hill's postal reforms. He introduced new, much lower postage rates based on weight, not distance, so the same value of stamp could be used for a letter to anywhere in the country. As a result the volume of mail delivered in the UK rose by 360 per cent over the next decade. Other countries were quick to adopt his ideas. The first stamps outside the UK were introduced in Switzerland in 1843.

Something to Think About . . .

Fraud has always been an issue for the postal services. Before prepayment, senders would use codes on the covers of letters so that the recipient could read the information and then refuse to accept the letter, thus avoiding the postage charge. Nowadays special papers, franking and printing techniques have overcome the re-use and counterfeiting of stamps. Luminescent inks, hidden images and holographic portraits are much harder to forge.

Penny Black
1 May 1840
Sir Rowland Hill, English social reformer, launches the adhesive Penny Black postage stamp, bought by the sender and stuck to the letter as proof of prepayment.

Penny Red

10 February 1841

It is hard to see a red postmark on a black stamp, so the colours are simply reversed: a black postmark on a Penny Red stamp, which remains in use until 1879.

5 Cent

1 July 1847

Benjamin Franklin, America's first Postmaster General, is the face on the first American postage stamp. Since then over 130 US Post Office stamp issues have carried his image.

20 Centimes

1 January 1849

Ceres, the Roman goddess of agriculture, adorns the first French stamp. Like the Penny Black, it is soon replaced (by the 20 Centime Orange) to make the black postmark more visible.

Anaesthetics_ 1842

William Edward Clarke (USA)

Many of the substances we now think of as anaesthetics started life in spiritual or recreational use, their medical uses first observed by accident or during prehistoric acts of ritual mutilation or sacrifice. Opium's analgesic benefits for example were completely overshadowed for a time by the reputation of seedy opium dens.

In the early 19th century, scientific demonstrations of the nature of ether and laughing gas became known as 'ether frolics', where audiences were encouraged to experience mind-altering sensations. While studying at Berkshire Medical College in Massachusetts, William Edward Clarke became the first person to administer ether as an anaesthetic to facilitate a surgical procedure. American physician Oliver Wendell Holmes coined the term 'anaesthesia' in 1846, from the Greek, meaning 'without sensation'.

Something to Think About ...

Cannabis, now almost exclusively a recreational drug, was one ingredient of a herbal general anaesthetic called mafeisan, administered by Chinese surgical genius Hua Tuo in the 2nd century AD. He performed over 150 successful breast-cancer operations using it, but destroyed the formula just before his death, setting medical progress back by centuries.

Alcohol

Traces on pottery made c7000 BC suggest alcohol is consumed in Neolithic China. Until the late 18th century it remains the only common sedative for pain relief.

Opium

Derived from the dried sap of the seed pods of the Opium Poppy (*Papaver somniferum*), opium is known to the Babylonians by 2225 BC and used in medicine by the ancient Egyptians. In 1527 Swiss botanist, physician and alchemist Paracelsus invents laudanum, tincture of opium.

Ether

Paracelsus also notes the analgesic qualities of ether, discovered in 1275 by Catalan alchemist Ramon Llull and first used in 1842 by American dental student W. E. Clarke.

Nitrous oxide

Discovered in 1772 by English scientist Joseph Priestley, nitrous oxide is laughing gas, and is used as a recreational drug until American dentist Horace Wells operates with it in 1844.

Morphine

German pharmacist Friedrich Sertürner first produces morphine from opium in 1806. Synthetic substitutes developed in the 20th century include methadone in 1946.

Chloroform

Chloroform, discovered independently in France, Germany and America around 1831, is first used anaesthetically by Scottish obstetrician James Young Simpson in 1847.

Cocaine

First identified in 1859, cocaine is introduced as a local anaesthetic by Austrian eye surgeon Karl Koller in 1884 at the suggestion of psychiatrist Sigmund Freud.

Barbital

In 1902 German chemists Hermann Fischer and Joseph von Mering discover the hypnotic properties of diethylbarbituric acid, barbital, the first barbiturate drug.

Fax Machine_1843

Alexander Bain (Scotland)

Fax machines only came into general public use in the 1980s and were soon made redundant by the arrival of the personal computer with e-mail and image scanners. But by then the basic technology for transmitting images had been around for over 100 years, predating the telephone.

Only large corporations and organizations, those for whom sending and receiving facsimiles was essential, could afford the expensive and complex early machinery. One famous early demonstration of the value of the fax came in 1907. A photograph of a man wanted in connection with a bank robbery in Stuttgart was circulated throughout Europe by the German police using Arthur Korn's Bildtelegraph system. The man was recognized and arrested in London.

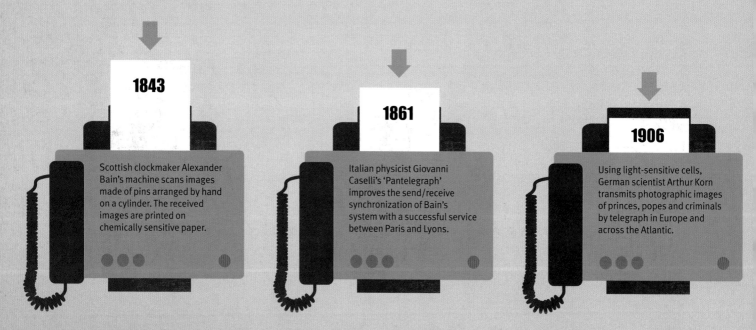

1843
Scottish clockmaker Alexander Bain's machine scans images made of pins arranged by hand on a cylinder. The received images are printed on chemically sensitive paper.

1861
Italian physicist Giovanni Caselli's 'Pantelegraph' improves the send/receive synchronization of Bain's system with a successful service between Paris and Lyons.

1906
Using light-sensitive cells, German scientist Arthur Korn transmits photographic images of princes, popes and criminals by telegraph in Europe and across the Atlantic.

Something to Think About . . .

In 1846 Alexander Bain adapted his electrochemical printing
process for use with standard telegraph alphabetical messages.
It could record nearly 300 words a minute compared with the mere
40 of which Samuel Morse's system was capable. Morse, however,
suppressed Bain's system with an injunction, ensuring that only
the lower speeds of Morse's own equipment would be used.
A case of Morse signalling 'SOS' – Send Only Slowly!

American designer Richard
Ranger sends the first
wireless fax image, a photo of
president Coolidge, from New
York to London with a system
still used by ocean shipping.

1924

Converting optical scans to
audio signals, Western Union
uses telephone lines to send fax
messages. The first, sent coast
to coast across America, features
images of Mickey Mouse.

1935

Xerox markets a new
generation of smaller,
simpler fax machines for small
business use, replacing the
earlier bulky and technically
complicated contraptions.

1966

Sewing Machine_1845

Elias Howe (USA)

Elias Howe built the first practical sewing machine, drawing on the innovations of many earlier inventors. He struggled to attract interest either in America or England, where he eventually sold his first machine. Returning to Massachusetts in 1849, he found that the idea of the sewing machine had finally taken off and that other manufacturers were infringing his patent.

In 1856, after several successful but lengthy lawsuits, the major manufacturers – Howe, Singer, and two others – agreed to pool their various patents. This saved them all a fortune in lawyers' fees and earned them all a further fortune in licensing agreements with other manufacturers, who had to pay $15 a machine in royalties to the four companies in the so-called Sewing Machine Combination.

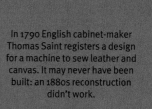

In 1790 English cabinet-maker Thomas Saint registers a design for a machine to sew leather and canvas. It may never have been built: an 1880s reconstruction didn't work.

Something to Think About . . .

Although not the inventor of the sewing machine, Isaac Singer was certainly an industrial pioneer. His company, I. M. Singer & Co., introduced the first domestic model in 1856 and was one of the first to adopt mass production by the use of interchangeable parts. It also offered the world's first hire-purchase agreements and triggered a sales boom when it began to accept trade-ins against new purchases.

1830

French tailor Barthélemy Thimonnier patents a machine for sewing seams in 1830. In 1841 rival tailors, afraid of being put out of work, destroy 80 machines in his factory.

1833

In 1833 American engineer Walter Hunt invents the lockstitch sewing machine. He doesn't solve early problems or patent it, fearing it will lead to unemployment.

1845

American inventor Elias Howe refines and, more importantly, patents Hunt's lockstitch approach in 1845. His big idea is a needle with the eye near the point, not the shank.

1851

In 1851 Isaac Merritt Singer adapts Howe's lockstitch, using a more reliable mechanism, with a straight vertical needle and a foot to hold the cloth in place.

Zip_1851

Elias Howe (USA)

It took over 80 years of design refinements for the zip to become the universal fastener it is today. In 1893, American sewing machine inventor Elias Howe's original idea from 1851, was revived by Chicago engineer Whitcomb Judson, for a friend who was having trouble lacing his shoes. In 1913 one of Judson's employees Gideon Sundbäck made further improvements including increasing the tooth count from four to ten per inch. The zip remained a fastening for footwear until it was adopted for children's clothing and trouser flies in the 1930s.

Something to Think About . . .

The word 'zipper' was coined by Goodrich & Co, a rubber-boot manufacturer, and one of the first businesses to use Sundbäck's new design, in around 1917. Sundbäck had given it a less snappy title – the Separable Fastener. Unsurprisingly, the name zipper stuck.

Laces
Early clothing was probably fastened with crude pins made from wood or bone, or with laced or sewn yarn. A pair of shoes with laces has been dated to 3500 BC.

Brooch
Pins became ornamental as well as functional. The earliest surviving brooches, for pinning cloaks around Bronze Age shoulders, are some 4,000 years old.

Press-studs
Horse halters used by the Chinese Terracotta Army of 210 BC incorporated snap fasteners. In modern times, German Herbert Bauer patented press-studs for trousers in 1885.

Buttons
Opposing rows of small brooches or buttons could be laced together. Thirteenth-century Germany saw the first buttonholes, when the fashion was for closer-fitting clothing.

Keeping it together

Buckles

The word 'buckle' is from the Latin word '*buccula*', and the earliest examples are found on the straps of Roman army helmets. Armour and tunics were also buckled on.

Hook and loop fastenings

A Swiss engineer, George de Mestral, inspired by his dog whose coat was covered in the burs of the burdock after a country walk, invented Velcro in 1941.

Zippers

The first recognizable zip was patented in 1851 but was not developed; its inventor Elias Howe was preoccupied with his earlier invention, the first modern sewing machine.

Elevator_ 1853

Elisha Otis (USA)

Human beings have always had the option of steps or ladders, but the development of elevators was driven by the need to raise products and materials, not people. The locking safety rollers devised in 1853 by American engineer Elisha Otis, originally to stop goods crashing down when a cable snapped, also made elevators safe for passengers for the first time. The invention of gearless traction mechanisms, by which most elevators still operate to this day, meant we could build taller buildings. Without the elevator, the skyscraper would not exist.

Something to Think About ...

The Otis Elevator Company was also a pioneer of that relative of the elevator, the escalator. Developing ideas first patented by Jesse Reno in 1892 and Charles Seeberger in 1895, the Otis Escalator won first prize at the Paris Exposition Universelle of 1900. If the elevator gave us skyscrapers, did the escalator give us shopping malls and mass transit hubs?

Going up – elevator firsts

Largest passenger elevator

The five 80-person elevators installed by Mitsubishi in Osaka's Umeda Hankyu office building in 2010 are the largest in the world. Each has a floor area of over 9.5 sq. m (102 sq. ft).

First safety elevator

The first safety elevator was installed by Elisha Otis into Eder Haughwout's five-storey Emporium at 488 Broadway, New York in 1857. Over 150 years later it is still there and still in working order.

First passenger elevator

The Flying Chair, built at Versailles in 1743, used counterweights and pulleys to raise Louis XV from the first floor to his lover's apartments on the second.

Tallest and fastest elevators

Officially opened in 2010, the Burj Khalifa (Khalifa Tower) in Dubai has 57 Otis elevators, including the world's tallest, which rises 504m (1,654ft); and the world's fastest, which travels at 64.4 km/h (40mph), 1,079m (3,540ft) per minute.

First gearless traction elevator

Developed by the Otis Elevator Company in 1903, the innovation meant elevator shafts were no longer limited in height by the length of cable it is possible to make.

Condom_ 1855

Charles Goodyear (USA)

Few inventions have aroused such moral outrage while at the same time delivering such obvious health benefits. Certainly in their original form they can hardly have been designed for sexual pleasure – early Japanese condoms were made of horn or tortoiseshell, while in ancient China they used oiled silk paper.

Animal intestines were the choice of Casanova and others, more comfortable but very expensive. They were therefore frequently washed and re-used. Such condoms were still available until the start of the 20th century, when cheap, mass-produced latex devices finally replaced them. The popularity of condoms declined in the 1960s with the advent of the female contraceptive pill and the decade of free love. But the benefits of barrier contraception were firmly reinforced by the emergence of AIDS and HIV in the 1980s.

Something to Think About . . .

The sale of condoms has been restricted at many times in many countries. Nineteenth-century state and federal laws resulted in American troops not being supplied with condoms during the First World War (unlike other armies). As a result, 70 per cent of American forces fighting in Europe were infected with sexually transmitted diseases by 1918.

The journey towards
'Durability, Reliability and Excellence'

1994

A stronger, thinner polyurethane condom appears on the market, another Durex innovation. It is not vulnerable to oil-based lubricants, which can rot latex and rubber.

C1925

In the USA, Youngs Rubber Company produces the first dipped latex condom, thinner and smoother than its rubber predecessor. In Europe, Durex move to latex in 1932.

1564

Italian physician Gabrielle Falloppio's book *De Morbo Gallico* contains advice on using linen condoms to prevent infection from 'the French disease' – syphilis.

1957

Durex, originally founded as the London Rubber Company, launches the first lubricated condom. The name reflects its aims of 'Durability, Reliability and Excellence'.

1855

After the invention of the vulcanization process in 1844, rubber tyre manufacturer Goodyear starts to market the first rubber condoms – with seams!

Plastic_1855

Alexander Parkes (England)

Plastic's infinitely variable qualities of lightweight flexibility, durability and non-toxicity have encouraged its spread into every area of human activity from leisure to healthcare. Its ability to take on any shape we can imagine has placed styling and design at the heart of even the most functional objects. The invention of plastics has transformed the world we live in.

Something to Think About . . .

You might be wearing old plastic bottles! Modern synthetic fleece clothing can be made of fibres spun from recycled PET (polyethylene terephthalate), commonly used for carbonated drinks bottles. From cool drinks to warm clothing – now *that's* recycling!

Parkesine (1856)

Alexander Parkes from Birmingham invents the first man-made material, Parkesine, a bioplastic derived from the cellulose found in plant cells. It is most widely used as an imitation ivory. It is an early form of celluloid that was later developed by American John Wesley Hyatt.

Bakelite (1907)

Belgian chemist Leo Baekeland invents the first completely synthetic plastic, Bakelite. Extremely hard, strong and cheap, it is widely used as a protective casing in objects such as telephones, clocks and radio sets, and in the manufacture of billiard balls.

Fantastic plastic!

Polystyrene (1930)

German chemical giant I. G. Farben begins to produce oil-based polystyrene, whose properties were first noted 100 years earlier. Today it is the material behind a vast range of products from packing foam to CD cases, from toys to airtight tubs.

Vinyl (1926)

Discovered accidentally in the 19th century, PVC (polyvinyl chloride) is first manufactured by the B. F. Goodrich Company based in Akron, Ohio. Brittle when pure, it is softened for use in pipes, banners, clothing and window frames.

Internal Combustion Engine_1859

Jean Joseph Étienne Lenoir (Belgium)

There had been sporadic attempts to devise internal combustion engines before the mid-19th century. What finally made them a viable alternative to the steam engine was the new commercial exploitation of oil and the production of cheap petroleum. Their reliance on fossil fuels has driven recent research towards more efficient design and even to alternative or hybrid sources of energy. But the Otto four-stroke cycle – intake of fuel, compression, combustion and exhaust – remains the basis of the engine, which has been used to power everything from lawnmowers to aircraft, and electrical generators to submarines.

Something to Think About . . .

Alessandro Volta, who invented the electric cell, also demonstrated an early form of internal combustion. In the 1780s he devised a toy pistol in which a mixture of hydrogen and oxygen was ignited by a spark. The expanding gases fired a cork from the end of the gun barrel in exactly the same way that fuel lit by a spark plug drives the pistons of an engine.

Motor Car_1885

Karl Benz is acknowledged as the inventor of the car. His invention used the internal combustion engine to power the first automobile. First built in Mannheim, Germany, in 1885, his three-wheeled, gasoline-powered car went into production in 1888. Modern life would never be the same again.

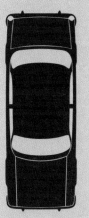

1859

In 1859 Belgian engineer Jean Joseph Étienne Lenoir adapts the steam engine, replacing steam with coal gas, which expands when ignited or combusted inside the piston cylinder.

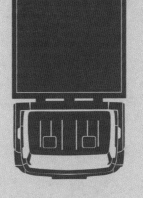

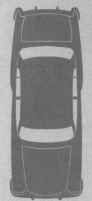

1897

In 1897 another German inventor, Rudolf Diesel, develops a more efficient form of internal combustion engine in which the fuel is ignited not by a spark but by the heat of compression.

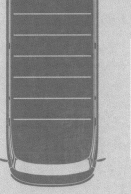

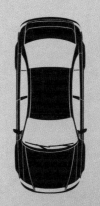

1861

German inventor Nikolaus Otto adapts Lenoir's design so that it can run on a mix of liquid fuel and air, and in 1861 builds the first practical four-stroke internal combustion engines.

1929

Felix Wankel, also from Germany, patents his simpler, smoother and more compact rotary engine in 1929. However, his four-stroke, three-chambered design is not produced until 1957.

Typewriter_1874

Christopher Sholes (USA)

Christopher Sholes' typewriter design was the first to achieve commercial success. It followed over 100 earlier attempts by him and others to develop a working machine. Many of them were noble efforts to enable the blind to write. Their eventual success was in the service of business, and the rise of the typing pool offered women a new, respectable alternative source of employment. It also led to the spread of shorthand skills. In a heyday spanning less than 100 years, the typewriter was central to changes in office working practice, and gave women new opportunities in society.

Something to Think About . . .

To save on costs, some early typewriters omitted 'unnecessary' characters from the keyboard. Instead of the numeral one, for example, typists were expected to use a lower case L; a capital O doubled as a zero; and an exclamation mark had to be typed with not one but three strokes – an apostrophe, a backspace, and a full stop!

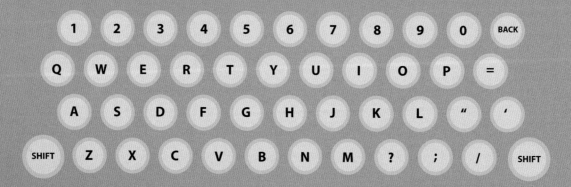

Where's the correction fluid?

(Key shifts in the development of the typewriter)

English inventor Henry Mill patents a 'machine for transcribing letters' in 1714. Although he probably built a working model, no details of it survive to the present day.

American newspaper editor and printer Christopher Latham Sholes sells the rights for his Sholes & Glidden Type-writer to Remington (a company that manufactured sewing machines and rifles) for a fee of $12,000. Sholes' backer, James Densmore, opts for a royalty deal and makes $1.5m.

Sholes devises the 'qwerty' keyboard layout for his typewriter shortly before the rights are sold to Remington; subsequent adjustments create the keyboard we know today. In the 1930s Dr Augustus Dvorak, a professor of experimental psychology at the University of Washington, produces an alternative that allows typing speeds up to 35 per cent faster, but it does not catch on.

In 1928 Remington turns down a long-term deal to produce the first electrical typewriter, invented by James Fields Smathers of Kansas City in 1914. Instead the contract is taken up by a subsidiary of General Motors, a company that becomes known as Electromatic Typewriters Inc., and is bought by IBM in 1933.

IBM coins the term 'word processor' in the 1960s for its electronic typewriters; but as computers begin to take over the company sells its typewriter operation to Lexmark in 1990.

Telephone_ 1876

Alexander Graham Bell (USA)

The telephone transformed mankind's ability to communicate and exchange information, accelerating social change and the development of ideas. It's hard to imagine an invention with a greater impact on how we live today. So it's little wonder that there are so many competing claims to have been its inventor.

In the end, two men filed patent claims to the same US patent office on the same day – 14 February 1876. Elisha Gray's was earlier, but Alexander Graham Bell's was processed sooner. Both men solved the problems of carrying the complex layers of sound frequencies found in human speech. But it was the Bell Telephone Company that prospered after the Western Union telegraph company turned down the chance to buy the patent for a mere $100,000, saying it would never catch on.

1854

Charles Bourseul (France)

'I publish a scheme capable of transmitting tones and vowels by electricity.'

1837

Charles Grafton Page (USA)

'I generate sound by passing current through wire between the poles of a magnet.'

Something to Think About . . .

The Bell Telephone Company fought off more than 600 rival claims in the courts. One of the strangest was from Daniel Drawbaugh, a backyard inventor from rural Pennsylvania who, in 1888, said he'd invented the phone back in 1867 using only a teacup as a transmitter. Case dismissed.

1831

Michael Faraday (England)

'I show that vibrations of metal objects can be converted into electrical impulses.'

1854

Antonio Meucci (Italy/USA)

'I invent an electromagnetic device I call a *telettrofono* but can't afford to pursue it.'

1860

Johann Philipp Reis (Germany)

'I transmit music and indistinct speech by diaphragm, needle and electrical contact.'

1864

Innocenzo Manzetti (Italy)

'I make a speaking telegraph which conveys vowels to the mouth of an automaton.'

1874

Poul la Cour (Denmark)

'I run audio telegraph trials on a telegraph line between Copenhagen and Jutland.'

Microphone_ 1876

Emile Berliner (USA)

Microphones convert sound waves into electrical current. The variations in the current reflect the changes in frequency of the sound waves. The medium by which that current varies is called the transducer, the critical element in any microphone's ability to capture sound faithfully.

Primitive microphones simply varied the distance between two electrical terminals, either along a solid metal strip or in an acid used as a liquid transducer. Berliner's carbon granules provided a more stable solution. New developments in microphone technology harness new mediums – fibre optics and lasers for example – in the quest to capture more perfectly that most direct device of communication, the human voice.

Converting sound to electricity

Carbon granules

In 1876 American inventor Emile Berliner designs the first recognizable microphone, for use in the Bell telephone. Sound waves captured by a flexible diaphragm compress carbon granules in the mouthpiece, varying their resistance to an electrical current.

Capacitors

E. C. Wente at Bell Labs develops the condenser microphone in 1916. The direct vibrations of the diaphragm, acting as one half of a capacitor, cause the required variations in current. It has a wide range of uses from phone calls to hi-fi recording.

Crystals

Instead of carbon granules, piezo-electric microphones use crystals that respond electrically to compression by sound waves. First used for naval sonar detection, the technology is also found in record-player cartridges and music instrument pick-ups.

Something to Think About . . .

Feedback occurs when a microphone picks up not only the
original sound but also the playback of that sound from
loudspeakers placed within earshot of the microphone.
The tone of the feedback depends on the resonant frequencies
of the equipment involved and the environment in which it
happens. Skilled rock musicians, for example, can control
feedback to good effect. Unskilled ones definitely can't!

Ribbons

Responding in the 1920s to radio's
need for better sound quality, German
physicists Schottky and Gerlach
replace the carbon granules with
a flexible ribbon between the poles
of a magnet. Sound waves move the
ribbon, inducing a variable current
within it.

Electromagnets

Dynamic microphones, invented by
E. C. Wente, use the same principle
as ribbon microphones, but an
induction coil attached to the
diaphragm moves within a magnetic
field. They are sturdy and slow to
feed back, making them ideal for
capturing live performances.

Electrets

Electrets are permanently charged
ferrous materials, and their use in
microphones was pioneered at Bell
Labs in 1962 by Gerhard Sessler
and Jim West. Such microphones
are reliable and cheap to produce,
and are found in most mobile
phones and computers.

Chapter 05.0

Phonograph
to Aeroplane
1877—1903

Phonograph_ 1877

Thomas Alva Edison (USA)

Having invented the phonograph, Thomas Edison was very quick to see the potential of sound recording. He foresaw its use in many ways which we now take for granted – voicemail, audio books, archive recordings of family members, office dictation, even speaking clocks.

Its greatest use, for most of the 20th century, was for entertainment. Early discs were made of hard rubber or more commonly brittle shellac. So-called unbreakable records, introduced in 1904, were cardboard discs with a coating of celluloid which contained the tracks. The first vinyl record was for a 1939 cigarette commercial, mailed out to radio stations.

Something to Think About . . .

The oldest sound recording is one that was never meant to be heard. In 2008, scientists scanned and digitized the wave forms recorded by a phonautograph (see opposite, top) on 9 April 1860, enabling them to be played back for the first time. They proved to be a version of the French folk song 'Au Clair de la Lune', recorded by the phonautograph's inventor, Édouard-Léon Scott de Martinville.

26 January 1857

The phonautograph, invented by French bookseller Édouard-Léon Scott de Martinville, records speech patterns. A stylus attached to a diaphragm makes marks on smoke-blackened glass, but it is a visual record only – no playback is possible.

30 April 1877

French poet Charles Cros's paleophone is never built, but he describes a track etched in metal by the recording needle, which can then be played back by a pick-up needle relaying the vibrations in the track to a diaphragm to re-create the original sound.

21 November 1877

Building on work with the telephone and the telegraph, serial American inventor Thomas Edison devises the phonograph, which can record and reproduce the human voice, through marks made in tinfoil by a needle following a spiral track on a cylinder.

June 1885

American engineers Chichester Bell and Charles Tainter develop an improvement of Edison's machine, the graphophone, using a wax-coated recording cylinder. The company formed to produce it eventually manufactures the first office Dictaphones.

May 1887

German-American inventor Emile Berliner's gramophone replaces Edison's cylinder with a horizontal zinc disc. After recording, the spiral track on it is etched with acid to prepare it for playback. The disc proves easier to mass-produce than the cylinder.

Light Bulb_1879

Thomas Edison (USA) and Joseph Swan (England)

After 75 years of experiments undertaken by scientists all over
the world, two men on either side of the Atlantic reached the same
conclusions, almost simultaneously. Joseph Swan in northern
England, and Thomas Edison in the USA, both filed patents in 1878
and demonstrated their incandescent light bulbs in 1879. Since then,
inert gases have replaced the original vacuum within the bulb and
tungsten filaments burn more brightly than their carbon predecessors.
Fluorescent lighting appeared in 1934, relying on properties of
luminescence (glowing) rather than incandescence (burning);
the same principle lies behind today's energy-efficient light bulbs.

Something to Think About . . .

The enduring image of a scientist at the moment of discovery
is the *light bulb* above his head, and we celebrate his *spark
of genius*, his *brilliance* and his *bright ideas*. The electric light
bulb is the instantly recognizable icon of invention.

Torch
400,000 years ago

This portable form of fire
allows man to explore his
environment in the dark,
whether it is a camp perimeter
at night or a cave system.

Camp fire
400,000 years ago

Early man learns how to
light fires, creating his
first controllable source
of heat and light.

The sun and moon

Although readily available,
these sources of light are
outside mankind's control,
are only effective in the open
air and are restricted by cloud
and position relative to Earth.

Oil lamp
10,000 BC

A dish containing a wick soaked in vegetable or fish oil provides a steady flame.

Candle
c300 BC

A wick wrapped in wax or whale fat can be stored and carried safely and easily.

Natural gas
4th century AD

The Chinese pipe natural gas to their homes through bamboo pipes to provide light.

Coal gas
1792

William Murdoch, an employee of engineer James Watt, is the first to use coal gas for domestic lighting.

Electric light
1802–09

In the early 19th century English inventor and chemist Sir Humphry Davy conducts a number of experiments on the effects of electricity on various chemical compounds. In 1809 he demonstrates his electric light (in fact an arc lamp) to the Royal Society.

Pneumatic Tyre_1888

John Boyd Dunlop (Scotland)

Not for nothing were the early bicycles called boneshakers. John Dunlop first devised his air-cushioned rubber tyres to ease the headaches suffered by his son while cycling on solid wooden wheels. By 1906 even aircraft had adopted the idea. From 1900 manufacturers experimented with cord reinforcements in the rubber walls of tyres – at first in the line of travel, then cross-ply (at an angle) and finally in 1947 radially (running across the tyre from lip to lip). Since the introduction of the first synthetic rubber tyre in 1937, other innovations in man-made materials have further improved the durability and efficiency of tyres in getting us from A to B quickly and comfortably.

Something to Think About . . .

Tyres get their name from their function. The first tyres were heated metal bands placed around wooden cartwheels. As the band cooled it shrunk, tightening around the wheel, thereby *tying* the rim and spokes together.

1891

Charles Terront wins the first ever long-distance bicycle race from Paris to Brest and back. He uses new removable pneumatic tyres devised by two French engineers, the Michelin brothers. Four years later they produce the first *pneus* for automobiles.

1888

Scottish vet John Boyd Dunlop straps inflated tubes of canvas and rubber to the wheels of his son's tricycle. The following year Ulster cyclist Willie Hume wins all his races on Dunlop tyres, conclusively demonstrating their advantages.

1847

Self-taught Scottish engineer Robert William Thompson demonstrates an inflatable rubber tube enclosed in leather, fitted to horse carriages in London's Hyde Park. A shortage of thin rubber persuades him to concentrate on solid rubber wheels instead.

1839

American industrialist Charles Goodyear stumbles on the process of vulcanizing rubber for strength and flexibility. It is used to make solid tyres for bicycles and coaches. English manufacturer Thomas Hancock devises a similar technique in 1844.

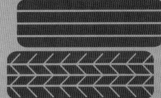
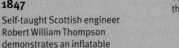

05.4 Electric Kettle_ 1891

Carpenter Electric Co. and Crompton & Co. (USA and England)

The first public electricity supply was introduced in England in 1881. As the use of electricity for light became more widespread, ingenious minds began to consider other ways of harnessing electric power for domestic use. There are records of numerous patents in the 1880s for electrical water-heating devices, although it is not clear whether any of them were actually made and sold. The earliest may have been English electrical engineer Mr Lane-Fox's egg boiler of 1886.

The earliest kettles were more novelty items than practical kitchen gadgets – they were very expensive and took 20 minutes to boil ½ litre (1 pint) of water. More efficient kettle design and better electrical engineering gradually reduced this over the next 20 years to less than ten minutes. Only in the 1920s, when Bulpitt & Sons took the step of immersing the element in the water, did the electric kettle really start to compete with the traditional stovetop kettle for speed and convenience. Modern kettles complete the task in around one minute.

Something to Think About . . .

The resistance-wire technology, which powered the first kettles, is still used in that other kitchen essential, the toaster. The earliest toasters only cooked one side; an automatic bread turner was introduced in 1913, and the first pop-up toaster appeared in 1919. Finally in 1925 the Waters Genter Company of America produced a toaster that cooked both sides at once, popped it up afterwards *and* switched itself off! Sliced bread followed three years later.

1955
Russell Hobbs, another English producer of electrical goods, unveils the first kettle with an automatic cut-out to eliminate the risk of boiling dry or electrical shock.

Boiling breakthroughs

1922
In England Bulpitt & Sons produce the first kettle with a heating element inside the water chamber – the heat source of earlier kettles was in a separate space below.

1891
The Carpenter Electric Co. in Minnesota, USA and Crompton & Co. in southern England, both launch an electric kettle, using resistance wire enamelled to the base.

Hip Replacement_1891

Themistocles Gluck (Germany)

Although Themistocles Gluck's innovative ivory joints provided only short-term relief from tuberculous bone decay, they proved a point. Surgery need not always be destructive – it could also be reconstructive. His work was a turning point, encouraging subsequent innovators to refine the problems of design and material, which Gluck's experiments were the first to address.

Charnley's breakthrough (see 1968, opposite) remains the basis for modern hip replacement prostheses. Insall (see 1972) has similarly provided the model for current knee procedures. Research in the 21st century concentrates on refining the mobility and durability of replacement joints, and on speeding post-operative recovery. New joints now routinely relieve the pain and transform the lives of millions of sufferers every year.

Something to Think About ...

The first successful shoulder replacement was undertaken in 1893 by French surgeon Jules Emile Péan, a great admirer of Themistocles Gluck's work. Péan had originally recommended amputation. But the patient, a Parisian waiter who carried food and drink for a living, understandably insisted he try first to save the arm.

1891
By 1891 German pioneer Themistocles Gluck has implanted several artificial joints of ivory in tuberculous hips, knees, elbows and wrists.

1925
In 1925, an American surgeon based in Boston, Marius Nygaard Smith-Petersen, casts a hollow glass hemisphere to fit like a socket over the femoral head.

1936
Vitallium, a new cobalt-chrome alloy devised in 1936, proves resistant to corrosion and therefore suitable for use in internal artificial joints.

1972

In 1972 Englishman John Insall devises a knee joint in which all three elements (upper and lower leg, and kneecap) are resurfaced.

1968

Charnley's student, Canadian orthopaedist Frank Gunston, applies the principles of metal-on-plastic to total knee replacement in 1968.

1958

From 1958, English surgeon Sir John Charnley develops the first total replacement hip procedure using stainless steel, Teflon and dental cement.

Hearing Aid_1892

Alonzo E. Miltimore (USA)

The ear trumpet, a sophisticated version of placing one's hand behind one's ear, was the dominant solution to hearing problems for at least 400 years. But sufferers of deafness wanted more discreet ways of coping with their impairment.

Telephone technology resulted, by the end of the 19th century, in the first body-worn hearing aids – a carbon microphone connected to a magnetic earpiece. Hearing aids also capitalized on subsequent electronic developments such as the triode, the piezoelectric microphone and, eventually, the transistor. Integrated circuits, microprocessors and electret microphones allowed for further miniaturization and greater control over the quality of relayed sound.

1892
New Yorker Alonzo Miltimore's 'magneto telephone for personal wear' is the first patented body-worn electrical aid. It is never produced, but many similar devices are.

1812
French surgeon Jean Marie Gaspard Itard invents a device held between a deaf person's teeth and against those of the speaker, using jaw and skull to reach the ear.

1624
The earliest reference to hearing aids in print is a guide to the correct use of an ear trumpet by French priest Jean Leurechon. Beethoven is among more recent users.

1836
English inventor Alphonse Webster patents an acoustic aid that sits behind the ear. In 1855 Edward Hyde designs the first American patent, a set of open headphones.

1957
French surgeon André Djourno performs an operation to insert the first cochlear implant, an electronic device designed to stimulate directly the inner ear of a profoundly deaf person.

1955
Dahlberg Inc. launch the D-10 Miracle-Ear, the first in-the-ear hearing aid, weighing 14g (½oz) and incorporating three transistors, a microphone and receiver.

Something to Think About . . .

The last manufacturer of ear trumpets to cease trading was F. C. Rein & Son of London, and did so in 1963. They had been in the hearing-aid business for some time. In 1819 they constructed an acoustic throne for the King of Portugal, incorporating hollow resonance chambers into the armrests and seat that supplied a discreet hearing tube leading to the King's ear.

Motion Pictures_1893

Thomas Alva Edison (USA)

French camera manufacturers the Lumière brothers may not have invented movies, but they certainly invented cinema. Prolific American inventor Thomas Alva Edison presented his kinetoscope, a moving-picture gadget for individual viewers, in 1893. But in 1895 the Lumières, Auguste and Louis, developed a projector which allowed many people to watch motion pictures together.

It was that shared experience that ensured the success of moving pictures. They became the entertainment choice of the masses, and competition to attract audiences was fierce. That competition raised the quality of film making to an art, while also driving it down to the lowest common denominator. It also fuelled the demand for novelty effects from synchronized sound to 3-D, some more successful than others.

First animation

From 1892 to 1900 French teacher Charles-Émile Reynaud presents 500-frame film sequences in public using his own praxinoscope, forerunner of the Lumière projector.

First 3-D movie

The Power of Love is shown in red-green anaglyph format in Los Angeles in 1922. The novelty effect will return in various formats in the 1950s, 1970s and 2000s.

First sound film

Following experimental shorts and newsreels in 1926, which include music, speech and sound effects, 1927's *The Jazz Singer* is the first feature-length talkie.

Something to Think About ...

Although cinema is a 20th-century phenomenon, the idea of projection has been around since the 15th century when Venetian engineer Giovanni Fontana threw images of demons on to walls with a magic lantern, an idea eagerly taken up by magicians and charlatans over the next 400 years.

First Cinerama™ experience

A wraparound visual feast using three projectors and a huge curved screen, Cinerama is launched in 1952, followed quickly by Todd-AO, VistaVision and others.

First IMAX™ screening

Canadian corporation IMAX projects an image about 22m x 16m (72ft x 52ft) at Expo '70 in Osaka, Japan, to show *Tiger Child*, bringing motion sickness to the movies.

First Sensurround™ rumble

The use of very low bass notes, whose vibrations were felt rather than heard, was a short-lived cinema novelty, first used to great effect on 1974's *Earthquake*.

Radio_1893

Nikola Tesla (USA)

It's all done with waves! Without radio there would be no television, no satellite navigation, no radar, no microwaves, no cellphones, no deep-space telescopes, arguably no pop-music industry or global celebrity. But the invention of radio was the inevitable result of mankind's overwhelming urge to communicate, to share knowledge and information. Radio's history is a unique collaboration of technical innovations, from Maxwell's first electromagnetic theories to Sony's first pocket transistor radio in 1955, from Marconi's first long-distance radio mast in 1895 to the ubiquitous satellite dishes of today.

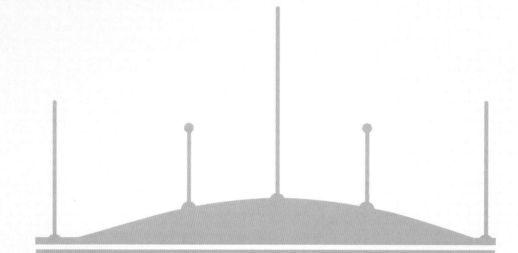

Something to Think About ...

At a huge electrical exhibition held in Madison Square Garden, New York in 1898, Nikola Tesla demonstrated the world's first radio-controlled vehicle, a boat which he called a teleautomaton. By 1915 he had conceived of unmanned warplanes, which did not become a practical reality until almost 50 years later, in Vietnam in 1964.

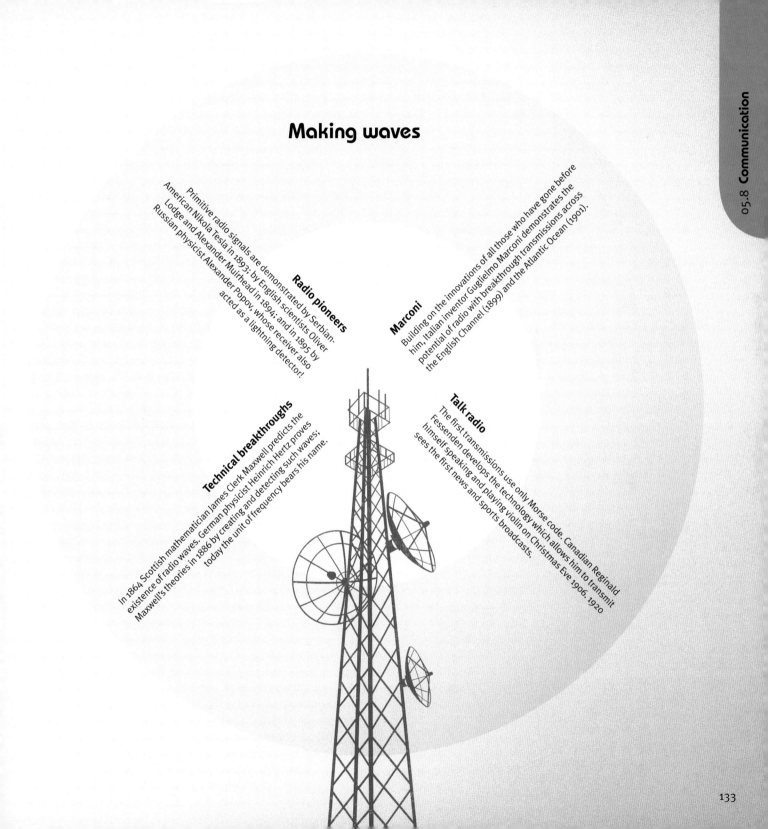

Making waves

Radio pioneers

Primitive radio signals are demonstrated by Serbian-American Nikola Tesla in 1893; by English scientists Oliver Lodge and Alexander Muirhead in 1894; and in 1895 by Russian physicist Alexander Popov, whose receiver also acted as a lightning detector!

Marconi

Building on the innovations of all those who have gone before him, Italian inventor Guglielmo Marconi demonstrates the potential of radio with breakthrough transmissions across the English Channel (1899) and the Atlantic Ocean (1901).

Technical breakthroughs

In 1864 Scottish mathematician James Clerk Maxwell predicts the existence of radio waves. German physicist Heinrich Hertz proves Maxwell's theories in 1886 by creating and detecting such waves; today the unit of frequency bears his name.

Talk radio

The first transmissions use only Morse code. Canadian Reginald Fessenden develops the technology which allows him to transmit himself speaking and playing violin on Christmas Eve 1906. 1920 sees the first news and sports broadcasts.

X-rays_1895

Wilhelm Conrad Röntgen (Germany)

X-rays had been observed as early as 1875 by several scientists conducting experiments with electrical-discharge tubes. But their principal interest was in cathode rays, and it fell to German physicist Wilhelm Röntgen to make a systematic study. He gave the name 'X' to the beams of electromagnetic radiation, simply because he did not know what they were. Although the expression 'X-ray' stuck in several countries, many others call them 'Röntgen rays' in his honour.

In 1895 Röntgen noticed that a fluorescent screen was glowing in the presence of these X-rays even though there was a barrier between them. This realization that they could penetrate solid materials inspired him to experiment, and the first ever X-ray photograph taken was of the bones in his wife's hand. Today X-rays have uses far beyond their original medical applications – for example, in airport security and deep-space observation. X-rays were also instrumental in the discovery of the double-helix structure of our DNA.

Something to Think About...

Medical diagnosis and treatment remain the uses of X-rays with which we are most familiar. Computer axial tomography – CAT-scanning – uses X-rays to view cross-sections of the body. Invented by English engineer Godfrey Hounsfield in 1971, it has transformed our understanding of how the body works. Some historians suggest that Hounsfield's research, conducted at EMI laboratories, was funded by the success of The Beatles, whose Apple record label was distributed by EMI's entertainment division.

The electromagnetic spectrum

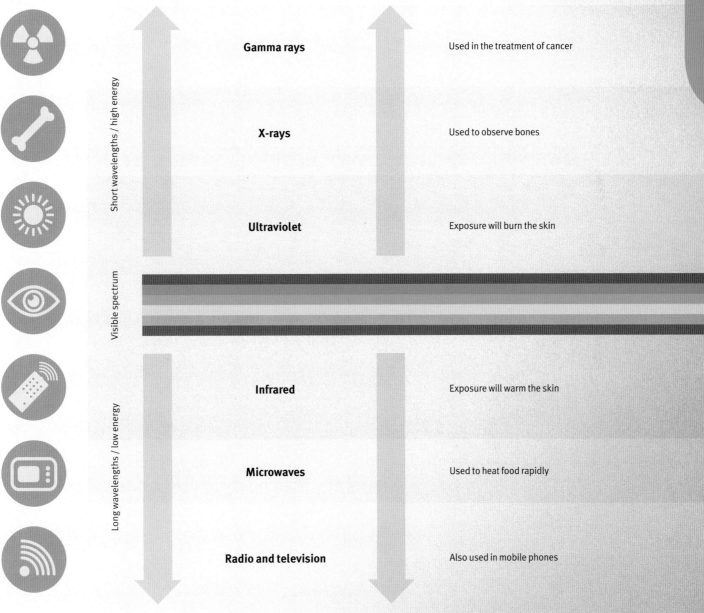

Short wavelengths / high energy

Gamma rays — Used in the treatment of cancer

X-rays — Used to observe bones

Ultraviolet — Exposure will burn the skin

Visible spectrum

Long wavelengths / low energy

Infrared — Exposure will warm the skin

Microwaves — Used to heat food rapidly

Radio and television — Also used in mobile phones

135

Washing Machine_1900

Unknown (USA)

Washing machines mechanize the age-old process of cleaning clothes by beating them on rocks beside a river. The ancient method used flowing water to carry the dirt away once it had been loosened by detergents and shaken or scrubbed on rocks. All this is now reproduced within the drum of your washing machine without you having to lift a finger to carry buckets of water or beat heavy, wet clothes for hours on end. No wonder that in a 2011 UK poll the washing machine was voted the most popular invention of all time. What is perhaps most remarkable is that the inventor of this indispensable household appliance remains unknown.

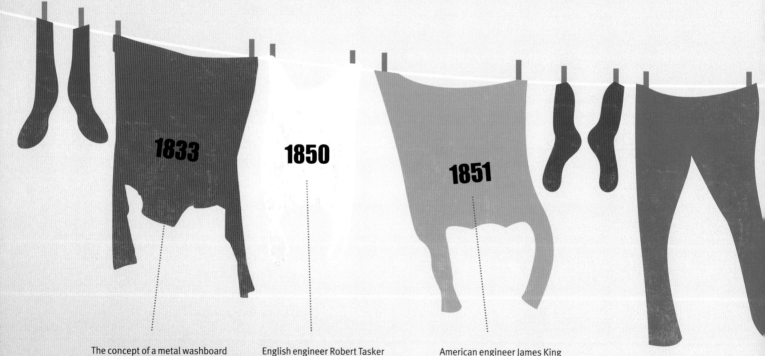

1833

1850

1851

The concept of a metal washboard is patented, appropriately, by American inventor Stephen Rust. Wooden ones have been used for centuries in northern Europe.

English engineer Robert Tasker designs a geared mangle for removing water from washed clothes, based on the box mangle already being used to press laundry.

American engineer James King patents a washing machine, the first to include the now-familiar drum container. But it is still hard work, being operated by hand.

Something to Think About . . .

The amount of water used by a washing machine raises environmental concerns. Improvements in efficiency have seen a drop from 150 litres (33 gallons) used by early machines to around 50 litres (11 gallons) today. It's all a great advance on the estimated 200 litres (44 gallons) or so of icy water a 19th-century laundry maid would have had to carry from the pump to the tub.

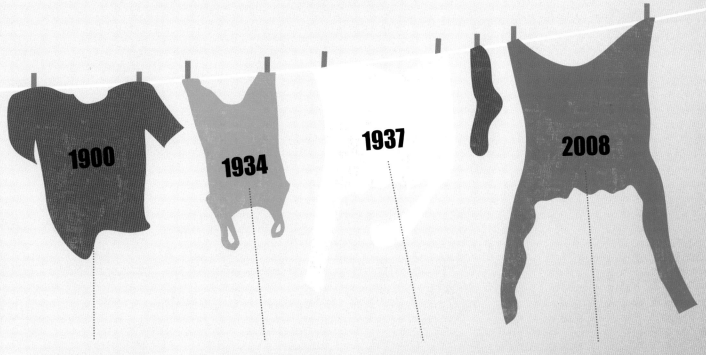

1900

1934

1937

2008

In the first decade of the 20th century several machines compete to be the first with an electric motor, finally taking much of the backache out of washday.

The first Laundromat opens in Fort Worth, Texas. In the UK, where public wash houses have been open since the 19th century, the first Launderette arrives in 1949.

In America, Bendix introduces the first automatic washing machine, so you no longer need to be present throughout the various stages of the washing cycle.

English chemist Stephen Burkinshaw develops a washing process that uses nylon beads and only 2 per cent of the water of traditional machines – about a cupful.

Vacuum Cleaner_1901

Hubert Cecil Booth (England)

The vacuum cleaner is an invention that could only have emerged in the affluent and temperate West. It became necessary because of our use of carpets, which cannot be swept like the hard, uncovered floors of poorer or warmer parts of the world.

The early devices of Thurman and Booth were large commercial machines, which were parked outside a building while hoses were run in through windows and doors to reach the dirty areas. James Spangler's domestic version was developed by the Hoover Company which offered an innovative ten-day trial period to boost sales.

As the use of servants declined, particularly after the First World War, every home needed a Hoover – and in Britain the name became synonymous with the act of vacuum cleaning, whichever manufacturer's machine was used.

Vacuuming up the dust of ages

1868 Chicago mechanic Ives McGaffey invents the first of many hand-operated vacuum cleaners. They are all awkward to use, powered by cranks, bellows, a foot

1898 In St Louis, Missouri, inventor John Thurman's horse-drawn, petrol-driven house-cleaning service blows compressed air into a confined space over the

1901 Cecil Booth, an English civil engineer, adapts Thurman's technique to suck instead of blow, and solves the problems of dust collection by

1907 Ohio janitor James Spangler designs a portable device using an electric fan, a pillowcase and – for the first time – a rotating, carpet-beating

1979 English inventor James Dyson adapts the industrial process of cyclonic separation, originally used in sawmills, for

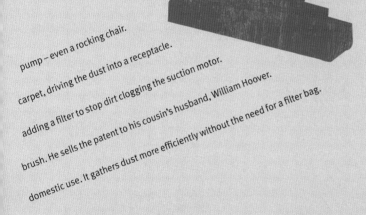

pump – even a rocking chair.

carpet, driving the dust into a receptacle.

adding a filter to stop dirt clogging the suction motor.

brush. He sells the patent to his cousin's husband, William Hoover.

domestic use. It gathers dust more efficiently without the need for a filter bag.

Something to Think About . . .

One of Cecil Booth's first cleaning contracts was a big one – the carpets in London's Westminster Abbey had become very soiled due to rehearsals for an important ceremony and had to be spotlessly clean. The ceremony? The coronation in 1901 of King Edward VII.

Aeroplane_1903

Wright Brothers (USA)

The history of the aeroplane is a twin tale of mankind's ingenuity and endurance. Long-distance pioneers such as Amelia Earhart, the first woman to fly across the Atlantic and Pacific oceans, had to conquer exhaustion and disorientation as well as the technical challenges of flight. Some paid an even greater price. Otto Lilienthal, a German whose glider experiments paved the way for the Wright brothers, was not the first or last to die trying to fly. His last words, in 1896, were 'Sacrifices must be made.'

But it is mankind's inventiveness we celebrate here. When George Cayley first defined the forces controlling flight – thrust and drag, lift and weight – flight became a real possibility. And when the Wright brothers made it a reality, they ushered in a new age. With our feet off the ground, we could start to look to the stars.

Something to Think About...

The Wright brothers built on the developments of many other pioneers. But they drew their greatest inspiration from gifts they received as children in 1878 – two rubber-powered toy helicopters devised by Alphonse Pénaud, a French inventor and admirer of George Cayley's work.

In 1977 American cyclist Bryan Allen makes the first human-powered flight, pedalling the 31.8kg (70lb) Gossamer Condor round a 1.6km (1 mile) figure-of-eight in California.

American engineers Orville and Wilbur Wright make the first manned, controlled, powered flight at Kitty Hawk in North Carolina in their biplane The Flyer, on 17 December 1903.

English engineer Sir George Cayley earns the title Father of Fixed-wing Flight. His 50 years of research culminate in the first gliders to carry a boy (1845) and a man (1853).

In 1907 French aviator Louis Blériot makes the first successful flight in a monoplane, and in 1909 wins £1,000 for the first powered flight across the English Channel.

How aviation took off

Italy first uses planes for military purposes, for reconnaissance and bombing during its war with Turkey, 1911–12. Planes will eventually change the way wars are fought.

Chapter 06.0

Helicopter to Teflon

1907—1938

Helicopter_1907

Paul Cornu (France)

Apart from da Vinci's fanciful drawing in the 15th century, it was not until the 1750s that inventors began to consider the real possibility of powered vertical flight. The main problem was one of stability, which early helicopters solved by using two counter-rotating blades. Sikorsky's breakthrough use of a single horizontal rotor in combination with a vertical tail rotor was the helicopter's coming of age. He resolved many practical issues of flight control, allowing the helicopter to take its place as an exceptionally versatile form of transport.

Surprisingly, both the largest and the smallest helicopters ever built have returned to the use of twin counter-rotors. In 1967 the Soviet Union built two prototypes of the massive Mil V-12, which had a lifting capacity of 105,000kg (231,483lb). At the other end of the scale is the 70kg (154lb) one-man GEN H-4, launched in Japan in 1998.

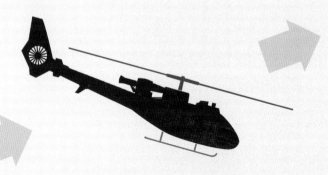

1861

In 1861 French engineer Gustave de Ponton d'Amécourt invents the word helicopter to describe his new steam-powered aluminium device. However, it fails to fly.

1480s

In the 1480s Leonardo da Vinci sketches a flying machine with a giant spiralling screw to provide lift. It is not a workable idea, as the blade tends to spin the craft.

We have lift-off

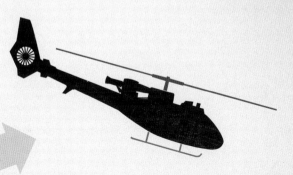

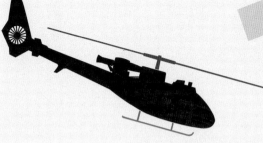

1907
French bicycle builder Paul Cornu achieves the first free helicopter flight when his machine raises him 30cm (1ft) off the ground for all of 20 seconds in 1907.

1940
Russian born aviation engineer Igor Sikorsky, who emigrated to the USA in 1919, patents a number of 'direct lift' aircraft during the 1930s. In 1940 his VS-300 helicopter flies untethered for the first time. With this achievement he solves all of the early helicopters' problems of steering and stability in the air.

Something to Think About ...

Since mankind first observed sycamore seeds spinning in the air, he has understood the principle of rotating blades. The oldest Chinese helicopter-like toys mimicked these seeds. Called 'take-tonbo' (literally 'bamboo dragonflies') they date to around 400 BC.

Teabag_ 1908

Thomas Sullivan (USA)

There are patents for teabags going back to 1903, but it was New York tea importer Thomas Sullivan who first popularized them in 1908 by sending out small samples of his blends, stitched into small silk bags. His clients didn't realize that they were meant to cut open the bag before putting the tea leaves into the teapot, and the teabag was born.

Silk gave way to gauze, and in 1930 to the heat-sealed paper sack, patented by American paper manufacturer William Hermanson. Square teabags arrived in 1944, and in 1964 Tetley Tea was the first company to introduce perforations, which speeded up infusion.

Something to Think About . . .

The British, a nation of tea-drinkers, resisted the idea of teabags, suspecting that the quality of tea would suffer if packed in porous material. It is true that poorer tea is used in bags; but teabags, which made up 5 per cent of British tea sales in the early 1960s, now account for around 95 per cent.

Tea strainer

Tea strainers have been in use as long as tea has been drunk – traditionally since 2737 BC. Tea made with loose tea is poured through a metal or ceramic filter. This allows the liquid to pass into the cup while trapping the leaves, which are bitter to taste.

Tea ball

The tea ball becomes fashionable in the early 19th century. Not only does it prevent loose leaves from entering the teacup; it can be removed from the teapot when the required strength has been reached, to prevent further brewing.

Tea press

The coffee press or cafetière has been in use since the 19th century. The tea press, developed in 1991 at the instigation of the British Tea Council, traps loose leaves within the pot when the tea is strong enough, restricting further infusion.

Teabag

The teabag has many advantages over the ball and the strainer – from the consumer's point of view it is ready-filled, cheap, easy and fast to use, and disposable. The great advantage to the manufacturer is that the quality of its contents is sealed from view.

06.3 **Hovercraft_ 1915**

Dagobert Müller (Austria)

Dagobert Müller's *Versuchsgleitboot*, designed for the Austro-Hungarian Navy, was actually a boat with hovercraft pretensions. But its use of a downward thrust of air to ease its passage across the waves supports its claim to be the first hovercraft. Christopher Cockerell's breakthrough was to add an inner wall below the craft. This greatly improved the efficiency of the air cushion on which the hovercraft relies, setting up air currents that recycled air instead of letting it escape under the skirt of the hovercraft. Today, hovercrafts are widely used by military and rescue services because of their capacity to carry loads at speed over difficult and varied terrain.

Something to Think About . . .

When the United Kingdom's royal consort Prince Philip took the controls during a demonstration of Cockerell's prototype hovercraft in 1959, he drove it so fast that the bow was dented by the waves. Cockerell refused to repair the damage (clearly visible in later photographs), christening it instead the Royal Dent.

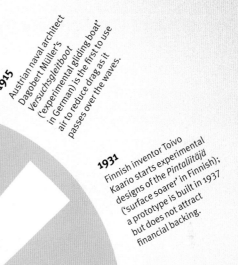

1915
Austrian naval architect Dagobert Müller's *Versuchsgleitboot* ('experimental gliding boat' in German) is the first to use air to reduce drag as it passes over the waves.

1931
Finnish inventor Toivo Kaario starts experimental designs of the *Pintaliitäjä* ('surface soarer' in Finnish); a prototype is built in 1937 but does not attract financial backing.

1935
Soviet engineer Vladimir Levkov produces around 20 experimental military hovercraft including the L-1, but technical problems and the Second World War halt further trials.

1959
English engineer Sir Christopher Cockerell finally launches the prototype of his hovercraft SRN1, first conceived in 1953. He also comes up with the word 'hovercraft'.

1988
The largest hovercraft in the world is the Russian Zubr class amphibian assault craft, which can carry up to 500 soldiers, or three main battle tanks. It is still in production.

Tank_1916

British Army (Battle of the Somme)

By 1915 the First World War had reached a stalemate. Under pressure from a population horrified by the waste of human life, commanders broke the deadlock with an armed and armoured vehicle whose caterpillar tracks could cross the cratered wastelands of No-man's Land and punch a hole in enemy lines. Although their impact was lessened by constant breakdowns, they proved their worth.

Between the Wars, there was much debate about the role of tanks within the armed forces. Originally conceived as support for infantry regiments, they developed by the Second World War into battle forces in their own right. Today they are central elements of any military land-based campaign.

Something to Think About . . .

Tanks get their name from an elaborate attempt to cover up their existence. They were to be called landships; but to throw any German spies off the scent, the word was spread that the vehicles were being developed as mobile water carriers for use in Mesopotamia. The engineers working on their construction took to calling them water tanks, and the name stuck.

The steady advance of the armoured tank

1917
At Cambrai, British Mark IV Tanks form part of the battle plan for the first time. They advance deep into enemy territory, with half the casualty rate of earlier attacks.

1916
Allied commanders use the Battle of the Somme to test the British Mk I Tank. Forty-nine are deployed, but are plagued by mechanical failure; only nine reach the German lines.

1915
Britain's Little Willie is the first of several wildly differing armoured prototypes to be constructed in the summer of 1915 by nations on both sides of the First World War.

1903
English science fiction writer H. G. Wells imagines an army of 30m- (100ft)- long armed tank-like vehicles in a trench-warfare context in his story *The Land Ironclads*.

1833
An anonymous 'Constant Reader' of the London United Services Gazette proposes a 'steam chariot of war' with an armoured boiler, to plough through enemy infantry.

1487
Also inspired by the tortoise, Leonardo da Vinci describes an armoured shell on wheels, driven by men protected inside, 'to take the place of elephants' in warfare.

AD 100
The 'testudo' (literally 'tortoise') deployed by the Roman army is an armoured formation designed to protect a body of soldiers as they advance through enemy lines. The shields are held in such a way that they present a protective wall to all sides.

Adhesive Tape_ 1923

Richard Drew (USA)

Although earlier patents for adhesive tape were registered by American inventor Henry Day (in 1848) and German pharmacist Paul Beiersdorf (in 1882), it was Richard Drew's innovations that finally caught on, thanks to the need to make do and mend during the financial hardships of the 1920s and 1930s.

Drew's willingness to develop different glues for use with different tape materials set the pattern for all later developments in adhesive tape. There are specialist tapes for every conceivable use, and nowadays it's hard to imagine life without it, in everything from surgery to plumbing, from labelling to gift-wrapping.

Something to Think About . . .

The first versions of masking tape only had glue on the outer edges, which is where Drew thought it would be needed. When the tape kept falling off the bodies of cars being painted, the painters accused the manufacturers of being ungenerous, or 'Scotch', with the amount of glue they were using. 3M changed its production process so that the whole of the back of the tape was coated, but the name Scotch Tape stuck!

1923 (Masking tape)
Richard Drew devises a 2in- (5cm-) wide tape backed with pressure-sensitive adhesive to help workers in the automotive industry who wanted a tidy edge to their painting work.

1930 (Cellophane tape)
Richard Drew adds adhesive to clear tape made of cellophane. It is popular because in the Depression of the 1930s people choose to repair things instead of replace them.

1942 (Duct tape)
Johnson & Johnson, which already manufactures medical tapes, develops a wide waterproof version for the US Army, to protect boxes of ammunition from humidity.

1946 (Electrical tape)
Researchers at 3M (which makes Scotch Tape) develop a glue for use on vinyl insulating tape – a component in the vinyl reacted with earlier tape glues, rendering them useless.

1960 (Surgical tape)
3M devises a tape for medical use, which is hypoallergenic, and allows the pores to breathe. Tape can also be used to apply certain medications absorbed by the skin.

Television_1925

John Logie Baird (Scotland)

John Logie Baird's demonstration of silhouette images in London in 1925 constituted the world's first television pictures. The BBC initially adopted his mechanical system but replaced it in 1937 with Marconi/EMI's electronic one.

Television works by converting pictures and sound into a signal, which is transmitted through the air, received by an aerial and decoded back into pictures and sound. The cathode-ray tube was a key component in the invention of television. It created the images and gave early sets their curved screens.

CCTV

Closed-circuit television, where pictures are transmitted to a specific place and a limited number of screens, was developed in Germany in 1942 to monitor the launch of V2 rockets. Its main use now is security and the prevention and detection of crime.

Colour television

Trialled in America in 1950, colour had replaced monochrome there by 1968. In the UK, the BBC did not begin general colour broadcasts until 1969.

Cable/Satellite

Cable was a 1940s American innovation while satellite was developed in the 1970s. These new means of reception would make subscription and specialist channels possible.

Something to Think About...

Viewers in Great Britain watch an average of 28 hours of
television each week, compared to 40 in the USA. By the age
of 70, the average American will have spent the equivalent
of eight years watching television.

Flat-screen television

Technological advances have made
the flat screen standard. Liquid
crystal display (LCD) consists of
millions of self-refreshing pixels.
In plasma TVs, the pixels are tiny
fluorescent plasma lamps.

Digital

Digital signals contain more
information than analogue,
producing better quality pictures
and making High Definition (HD)
television possible in the 2000s.

Magnetic Recording Tape_1928

Fritz Pfleumer (Germany)

Magnetic recording of sound was first achieved in 1928 by German-Austrian engineer Fritz Pfleumer who covered a long, narrow strip of plastic with a thin magnetizable coating. He granted the rights to use his invention to German electronics firm AEG which developed the first tape recorder by incorporating tape heads. Despite the advance of digital formats, tape is still commonly used in recording studios.

Reel to reel

Pioneered in Germany in the 1930s, reel-to-reel machines go on to be used extensively in the recording studio and the home.

Multi-track recording

First used to create a stereo effect with two tracks in 1943, the addition of more tracks allows instruments to be recorded separately, a process known as overdubbing.

Video Cassette_1951

In 1951, magnetic tape was adapted to record visual images, creating videotape. The video cassette was developed in 1969, paving the way for home videos in the 1980s.

1962

Compact cassette

Philips develops the cassette using narrower tape enclosed in a plastic case. Within a decade of its introduction in 1962 the cassette has largely replaced reel to reel in domestic use.

1987

1970s

Dolby noise reduction

In the 1970s, Dolby Laboratories create several systems to reduce tape hiss for both the recording industry and the cinema.

Digital audio tape

Launched in 1987, Sony's digital format known as DAT is intended as a replacement for the cassette but fails in the marketplace.

Something to Think About . . .

Before the invention of magnetic recording tape, radio programmes had to be recorded live and could not be kept for posterity. Similarly, gramophone recordings had to be completed in one take.

Jet Engine_1930

Frank Whittle (England)

The idea at the core of the jet engine, a backward thrust of gas or liquid delivering forward motion, has been around since about AD 100; Hero of Alexandria built a parlour curiosity which demonstrated that jets of steam could spin a sphere on its axis. Chinese armies used the same principle in the 13th century to fire rockets at their enemies. In the 1930s several companies were racing to fly the first jet plane. Hans von Ohain later declared that if the RAF had accepted Frank Whittle's ideas in 1930 there would have been no Second World War: Hitler believed that air superiority was vital, and would have had nothing to match Britain's early lead.

Something to Think About . . .

Jets aren't just for planes. Modified jet engines also turn up as gas turbines in power plants, pumping stations and ships' engine rooms. The land speed record, 1,228km/h (763mph), is currently held by Thrust SSC, a jet-propelled car, which was also the first to break the sound barrier.

1930 English air mechanic Frank Whittle presents his plans for a jet engine to the RAF. The RAF rejects the idea as unworkable, leaving Whittle free to patent it for himself.

1936 German physicist Hans von Ohain, unaware of Whittle's work, takes his ideas of jet propulsion to German aircraft manufacturer Ernst Heinkel. Heinkel welcomes him.

1939 Days before the outbreak of the Second World War, the Heinkel He178 is the first plane to fly entirely under jet power, propelled by von Ohain's HeS-3 turbojet engines.

1941 Jet engines are being developed in Italy, Russia and America, and the RAF finally flies a test plane, the Gloster E28/39, powered by Whittle's W-1 engines.

1978 Whittle and von Ohain, both now living in America, meet, become firm friends and travel round the USA on joint lecture tours, speaking about their shared invention.

Radar_ 1935

Robert Watson-Watt (Scotland)

In 1912 the unsinkable *Titanic* sank, and war was looming in Europe. The twin threats of icebergs and submarines prompted the first serious research into underwater detection by sound waves. Early radio pioneers including Marconi and Tesla foresaw the application of radio waves for the same purpose in the air.

The ability to detect approaching enemies made radar a focus of military research between the Wars, and Britain's lead in the field gave it a winning advantage against German air attacks. In times of peace, radar has also found useful applications in air and marine safety, in the monitoring of weather conditions and the observation of deep space. Ground-penetrating radar has helped us understand our own geology and is increasingly a tool for archaeological fieldwork.

Something to Think About ...

What's in a name? RADAR stands for RAdio Detection And Ranging. SONAR, SOund Navigation And Ranging, was a term coined in the USA long after the invention of sound wave detection, simply to match RADAR. ASDIC, the original term for SONAR, never actually meant anything – it was a code word used by Britain's ASD or Anti-Submarine Department, to conceal its interest in sound waves or sonics.

1915
Canadian engineer Reginald
Fessenden designs crude
echolocation devices for ten
British submarines built in
Montreal. They detect range
but not direction.

1918
Another Canadian, physicist
Robert Boyle, heads the
secret British project that
develops ASDIC, using
ultrasonic quartz crystals to
identify direction and range.

1954
The radar gun, much used for
speed-limit enforcement and
sports statistics, is invented by
Illinois scientist Bryce K. Brown
and tested on the busy streets
of Chicago.

1936
The US Navy and Army,
working independently of one
another, conduct successful
radar trials. During the Second
World War, America and Britain
pool their research.

1935
Scottish meteorologist Robert
Watson-Watt demonstrates the
radar system he has developed
for the British Air Ministry. Army
and Navy systems follow by 1937.

Nylon_ 1935

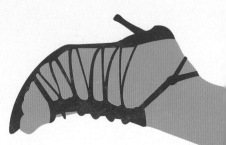

Wallace Carothers (USA)

Early nylon products included tents and parachutes in support of America's war with Japan, prompting rumours that 'Nylon' stood for 'Now You've Lost, Old Nippon'. It was only after the Second World War that its full potential began to become apparent. It is far more than a fine filament for textile production. Its strength and heat resistance make it a useful material for machine components, or in combination with other materials, such as glass or carbon fibre.

Disposal of used nylon items is a growing problem. It decays very slowly in landfill, and incineration releases toxic hydrogen cyanide gas. But in 1969 this very resilience made nylon the perfect material for the flag which Neil Armstrong planted on the moon.

Something to Think About ...

As the war effort had first claim on the supply of nylon, 'nylons' became scarce items, sold by black marketeers at four times their official price. When the stockings were made available again, Macy's in New York sold its entire stock of 50,000 pairs in just six hours.

1802

French immigrant Eleuthère Irénée du Pont sets up a gunpowder mill in Wilmington, Delaware; during the Civil War he supplies half the powder needs of the Union army.

1903

Now generally known as DuPont, the company sets up experimental chemistry laboratories in an effort to move away from its original business. Early successes include the development of cellulose.

One step ahead

1940
72,000 pairs of artificial silk stockings, 'nylons', are bought on their first day on sale.

1938
The first product to use the new material is not an item of clothing but a toothbrush.

1930
DuPont chemist Wallace Carothers is hired to head research into synthetic materials, and discovers Neoprene, the world's first synthetic rubber and the first of his 50 patents.

1935
Carothers invents nylon, an artificial silk, designed to relieve America's dependency on Japanese imports as the relationship between the two nations deteriorates.

163

Ballpoint Pen_1938

László Biro (Hungary)

Several inventors had tried to come up with a workable ballpoint pen before Hungarian László Biro filed his patent in 1938. Newspaper editor Biro noticed that the ink used on newsprint dried much more quickly than fountain-pen ink. He improved the ink-flow problems of his predecessors by using a small ball-bearing as the nib, drawing ink from the pen's reservoir. Before the ballpoint, writing was slow and messy; fountain pens often leaked and 'stick' pens needed to be dipped regularly into the inkwell. (See pages 50–51 for details of the ballpoint's predecessors, the pencil and the fountain pen.)

Something to Think About ...

The Space Pen was developed by Paul C. Fisher in 1965 and sold to NASA – not the other way around. The pen is capable of writing upside down, in zero gravity and in extreme conditions by using pressurized ink flow.

* **Flying High   1944**

   The British Royal Air Force is one of the first organizations to use Biro's design because his pens work at high altitudes. His name subsequently becomes synonymous with ballpoint pens in Great Britain.

* **Pen Wars   1945**

   Eversharp license Biro's pen in America but an unauthorized version, dubbed Reynolds' Rocket, is the first to appear in the shops.

* **Gimbels   1945**

   When ballpoints go on sale to the public for the first time in 1945, 5,000 people flock to Gimbels department store in New York where the entire consignment of 10,000 pens is sold in a day.

* **BiC   1945**

   Biro licenses his invention to Marcel Bich. Trading as BiC, the French company becomes the world market leader, refining the ballpoint and mass-producing it to sell cheaply.

* **Papermate   1949**

   Patrick J. Frawley solves the problems of smearing by developing a new ink which helps his Papermate pens sell in millions.

Teflon_ 1938

Roy Plunkett (USA)

Polytetrafluoroethylene (Teflon to you and me) was an accidental discovery. American chemist Roy Plunkett worked in the research laboratories of DuPont, the company that invented nylon in 1935. DuPont was looking for a safe coolant for use in domestic refrigerators, to replace the less-than-safe propane, ammonia and sulphur dioxide commonly used for the purpose at the time.

Plunkett came into the lab one morning to find that the compressed gas tanks with which he had been experimenting had seized up. Attempts to unblock the valve failed, and when Plunkett removed the valve he found not gas but a coating of slippery white flakes of an unknown substance inside. It was a polymer whose creation had been catalyzed by the iron walls of the tank under pressure. Plunkett's tests proved the substance's high resistance to corrosion, stress, temperature, acid and just about anything else. It was patented in 1941 and the name Teflon was invented for it in 1945.

Something to Think About . . .

In 1954 a French engineer and amateur angler Marc Grégoire used Teflon to help his fishing line glide more easily across the water. It was his wife who suggested that, if Teflon really was that slippery, it might be useful as a lining for her aluminium cooking pots. This first meeting of *Tef*lon and *Al*uminium gave the world both Teflon cookware and what became TEFAL, the company formed by Grégoire to produce it.

Uses of Teflon

Clothing
Gore-Tex, the waterproof but breathable fabric from which outdoor clothing is made, is an expanded polytetrafluoroethylene, a derivative of Teflon.

Bombs
Teflon was first used as a sealant on containers of corrosive uranium hexafluoride during America's development of the atomic bomb.

Cosmetics
Some nail varnishes include Teflon to make them harder wearing in the constant fight against chipped manicures.

Upholstery
Stain-resistant carpets owe their qualities to a coating of Teflon, which repels anything spilled on to it.

Bullets
Some armour-piercing bullets are coated with Teflon to ease their high-powered passage down the gun barrel.

Transport
Notice how your windscreen wipers don't squeak as much as they used to? It's because their blades have a thin coating of Teflon to help them slide more easily.

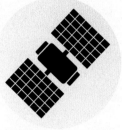

Chapter 07.0

Photocopier to Laser

1939 — 1960

Photocopier_1939

Chester Carlson (USA)

A New York patent attorney, Chester Carlson first devised a photographic method of copying documents in 1939. Previously, copies were made with carbon paper or manual duplicating machines. At first, Carlson was unable to sell his invention, which he called 'electrophotography'. He developed it under the auspices of the non-profit Battelle Memorial Institute which granted a manufacturing licence to the Haloid Corporation in 1947. The copier took on a new name, Xerox, which was soon applied to the company too. The first Xerox machine was built in 1949, after which the photocopier became a familiar feature of the modern office.

'Xeroxing' has become synonymous with 'photocopying' but the Xerox Corporation has fought hard to prevent its trade name from becoming a generic term. The company generally requests that dictionaries modify their entries to reflect this.

Duplicating machines

After the invention of the photocopier, the duplicator remains in use as an inexpensive form of document reproduction. The ditto machine, with its distinctive purple ink, is particularly popular in schools.

Colour photocopying

Available from 1968, colour photocopiers were initially viewed with suspicion because of concerns over counterfeit currency.

Digital technology

Combined printer/scanners start to replace photocopiers in the 2000s, with the ultimate aim of creating the paperless office.

Something to Think About . . .

Inside the copier is a drum charged with static electricity that attracts particles of a very fine powder known as toner. The inserted sheet of paper is charged with static electricity and pulls the toner off the drum. The toner is heat sensitive, so the loose particles fuse to the paper, thus creating an image.

Forensic identification
Documents can be traced back to the model and make of copier, sometimes to a particular machine, by analyzing imperfections in the copies.

Copyright
The increased use of photocopying leads to concerns over copyright infringement, and most countries adopt a fair-use policy for research purposes.

Computer_ 1941

Konrad Zuse (Germany)

The Machine Age saw the invention of contraptions for increasingly ambitious manufacturing tasks, machines that had to be constructed to ever-greater degrees of accuracy. As a result we were faced with significantly more complex mathematical calculations. The first computers were attempts to mechanize these calculations. They came from both industrialists and scientists, all seeking to remove human error from the process.

The ability to process a large volume of calculations and data created the need to store and transfer both the results and the programmes that generated them. Hollerith's 1890s punched cards were widely used as late as the 1960s, when IBM developed the floppy disk. The diskette was launched in 1971, surviving until the early 2000s. As computers became more central to everyday life, recordable CDs and flash drives became necessary to satisfy our ever-growing demands for memory.

1837
From 1837 English mathematician Charles Babbage adopts the idea of punched cards for his steam-powered Analytical Engine, the world's first general-purpose computer.

1890
Faced with the mass of information to be processed from the 1890 census, American civil servant Herman Hollerith devises a machine that uses punched cards not only to control systems but to enter raw data.

Something to Think About ...

The English poet Lord Byron has an unlikely connection with the early development of computers. His daughter Ada Lovelace was a talented mathematician who corresponded with Charles Babbage and devised a series of punched cards for use with his machine. She is widely recognized, therefore, as the first computer programmer.

1941
German pioneer Konrad Zuse builds a series of electromechanical machines including, in 1941, the Z3 – the world's first working electric programmable computer, using a simple binary system and information stored on punched film.

1943
In 1943 a British team led by Tommy Flowers builds the first of ten Colossus machines, the world's first digital computers, used to decode intercepted German communications and kept secret until the 1970s.

Suntan Lotion_1944

Benjamin Green (USA)

American airman and pharmacist Benjamin Green formulated the first suntan lotion in 1944 after witnessing the harmful effects of prolonged exposure to the sun on soldiers in the Pacific. His original petroleum-jelly-like formula contained a simple blocking agent but was sticky and unpleasant. Purchased and refined by pharmaceutical giant Merck, it was successfully marketed as Coppertone in the 1950s. Suntan lotion, also known as sunscreen, is a mix of organic and chemical compounds that absorb ultraviolet radiation from the sun's rays while stimulating melanin, the pigment that causes the skin to darken.

Sun Protection Factor (SPF)

In 1938, Austrian chemist Franz Greiter develops an early form of sunscreen and, in 1962, devises the Sun Protection Factor scale to which all suntan lotions must now conform.

UVA/UVB

Two types of ultraviolet radiation affect the human skin. UVA is less likely to cause sunburn but penetrates deeply and is more dangerous. UVB is the primary cause of reddening and tanning.

Something to Think About . . .

Moderate exposure to ultraviolet radiation from the sun is beneficial to health as it produces Vitamin D3, which supports the immune system. Too much sunbathing can cause skin cancer, premature ageing and cataracts.

Sunblock

Sunblock deflects the sun's radiation and cannot be used for tanning.

Indoor tanning lotion

Using indoor tanning lotion after sunbathing stimulates the production of melanin.

Fake tans

A tanning effect can be achieved by lotions that cause temporary browning of the skin by stimulating a chemical reaction or by those containing a bronzing dye.

175

Microwave Oven_1945

Percy Spencer (USA)

The microwave oven was discovered by accident in 1945 by American engineer Percy Spencer. While building a magnetron to generate microwaves for a radar set, he noticed that a chocolate bar in his pocket had started to melt because of the microwave energy. He adapted the radar technology by feeding the microwaves into a box to create the prototype oven. Spencer's employers, Raytheon, patented the invention and tested it in a Boston restaurant before constructing the first microwave oven to be sold commercially.

Microwaves are very high frequency radio waves that cause the water molecules in food to rotate, causing friction, which in turn produces heat and cooks the food quickly. In the 21st century we take its speed and convenience for granted.

Something to Think About . . .

The first commercial microwave oven was as big as a freezer. At almost 1.8m (6ft) tall, it weighed 340kg (750lb) and consumed 3kW of power, three times as much as today's models, and had to be water-cooled. Not exactly what you'd expect to find in your kitchen ...

1947
American company Raytheon launches the Radarange, the first commercial microwave oven but it is large, cumbersome and very expensive.

1955
When it licenses microwave technology from Raytheon, Tappan Appliances becomes the first company to produce microwave ovens for home use, but these too are not a success.

1959
The first Raytheon Radarange microwave oven is installed (and remains) on board the nuclear-powered cargo-passenger ship, NS *Savannah*.

1967
Raytheon, now the Amana Corporation, introduces the first popular line of domestic microwave ovens.

1970
Litton Industries develops a microwave with a now-familiar short, wide shape. Unlike its predecessors, this oven could survive being accidentally used while empty.

1975
In the USA sales of microwave ovens exceed those of traditional gas cookers for the first time.

Atomic Bomb_1945

Robert Oppenheimer (USA)

An atom bomb harnesses one of two ways of releasing huge amounts of energy – fission (splitting the nuclear particles of atoms apart), and fusion (forcing them together).

Fission is easier to achieve, and all nuclear power stations use this method to generate nuclear energy. In fission bombs, the use of a conventional explosive is enough to start a nuclear chain reaction. Fission bombs usually split atoms of uranium-235, the most readily available radioactive material.

Fusion requires very high temperatures – of the kind generated inside stars like the sun, or in nuclear explosions. Thermonuclear or fusion bombs therefore use the power of a fission explosion to set off a fusion reaction. Hydrogen bombs are fusion bombs using deuterium or tritium, which are heavy isotopes of hydrogen.

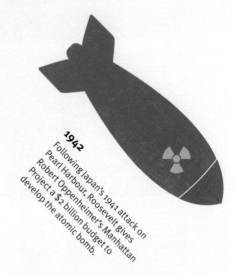

1942
Following Japan's 1941 attack on Pearl Harbour, Roosevelt gives Robert Oppenheimer's Manhattan Project a $2 billion budget to develop the atomic bomb.

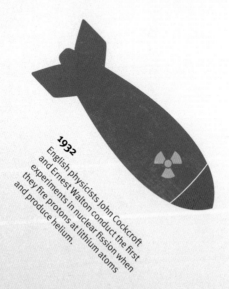

1932
English physicists John Cockcroft and Ernest Walton conduct the first experiments in nuclear fission when they fire protons at lithium atoms and produce helium.

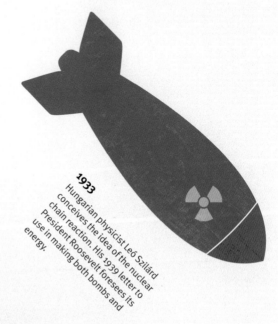

1933
Hungarian physicist Leó Szilárd conceives the idea of the nuclear chain reaction. His 1939 letter to President Roosevelt foresees its use in making both bombs and energy.

Something to Think About . . .

Although there have been thousands of nuclear tests since 1945, the bombs that fell on Hiroshima and Nagasaki are still the only nuclear bombs to have been used in wartime. Enola Gay, the B-29 superbomber that carried the Hiroshima device, was named after the mother of Paul Tibbets, the pilot who flew the mission.

1945
After a test explosion in New Mexico, directed by Oppenheimer, the USA uses two atomic bombs against Japan. On 6 August 'Little Boy' is dropped on Hiroshima and three days later, 'Fat Man' falls on Nagasaki: 150,000 people are killed instantly.

Transistor_ 1947

**John Bardeen, Walter Brattain
& William Shockley of Bell Labs (USA)**

Transistors are made possible by the nature of semiconductors – materials whose electrical conductivity changes under the influence of external energy. A transistor is a triode, a sort of sandwich of two kinds of semiconductor. It has two functions in electronic circuitry. It can act as an amplifier, in which a weak input signal is boosted to become a stronger output. Alternatively, a signal sent to the sandwich 'filling' can turn it into a switch, stopping current from passing between the two pieces of 'bread'.

These two simple uses – amplifying and switching – have put transistors at the heart of just about every aspect of 21st-century electronics. Before transistors, triodes were slow, expensive, bulky and fragile – delicate constructions of glass, inside which current passed from one terminal to another through a vacuum. The small, efficient, robust transistor was cheap to produce. With it, circuitry could be constructed in a compact, solid state – no unreliable mechanical switches or tubes. It gave us mobility, miniaturization and the Electronic Age.

Something to Think About ...

The Bell Labs team of Bardeen, Brattain and Shockley won the Nobel Prize in physics for their breakthrough. Bardeen won it again in 1972 as part of the team that developed the Bardeen-Cooper-Schrieffer theory of superconductivity. He remains the only person ever to achieve the Nobel physics double.

Dial to dial: from the telephone to the transistor radio

Hungarian physicist Julius Lilienfeld files a Canadian patent for a triode using semiconductors, but without supporting research or working models.

Using the semiconductor germanium, a team at Bell Labs, the research arm of American company AT&T, builds on Lilienfeld's theories and constructs the first working transistor.

American inventor Lee De Forest develops the vacuum-tube triode used to boost long-distance phone signals across the North American continent.

The Regency TR-1, made in Indianapolis, is the world's first mass-produced transistor radio. It uses a 22.5v battery and sells at $49.95 – around $400 allowing for inflation.

1925

1947

1906

1954

Ultrasound Imaging_1949

Dr John Wild (USA)

Sometimes known as sonography, ultrasound imaging is a medical technique that uses high-frequency sound waves to inspect the body's soft tissues and internal organs. Sound waves are sent into a particular area and the echoes that bounce back create an image. It is similar to shipping navigation system, SONAR. The technique was pioneered by Dr John Wild while carrying out research at the University of Minnesota. Magnetic Resonance Imaging (MRI) was developed in Scotland and first tested on humans in 1977. The patient is conveyed into a large chamber that generates a magnetic field to produce a picture of what is happening inside the body.

Uses of ultrasound

Foetal ultrasound
One of the most widespread uses of ultrasound is to monitor the progress of the foetus during pregnancy.

Echocardiogram (ECG)
An ECG uses ultrasound to look at the heart and detect any irregularities.

Something to Think About . . .

Ultrasound devices are small and often portable, making the technique more cost-effective than MRI. However, ultrasound is not as effective through bone. An MRI scan is static, whereas 4-D ultrasound can provide a moving image. Both are safer than X-rays, as no radiation is emitted.

Doppler ultrasound

This is a special ultrasound technique that evaluates the blood flow through blood vessels, including the major veins and arteries.

Ultrasound biopsy

During a biopsy a tissue sample is removed for analysis. Ultrasound guides the needle to the appropriate place in the body.

MRI scans

Magnetic Resonance Imaging (MRI) is typically used to detect damage to the joints and muscles, tumours inside an organ, and hematomas (bruising) in the brain.

Credit Card_1950

Ralph Schneider, Matty Simmons & Frank McNamara (USA)

The first credit card to be accepted by multiple retail outlets was the brainchild of the co-founders of the Diners Club, Frank McNamara, Ralph Schneider and Matty Simmons, in 1950. Diners Club was the world's first independent credit-card company, issuing cards for travel and entertainment purposes. The billing arrangements resembled a charge card, as the outstanding balance was to be paid in full at the end of each month. In 1958, BankAmerica introduced the first credit payment system. The technology that made it all possible was a magnetic strip on the back of the card, which was decoded by a reader.

FIRST CREDIT CARD

The first credit-card charge is made by Schneider, Simmons and McNamara in a restaurant in the Empire State Building, New York.

1950 XXXXXX XXXX

Something to Think About ...

Credit cards quickly became popular in Britain, America and Canada but take-up has been slower in more cash-orientated countries. Germany, France and Switzerland tend to favour debit cards, whilst in Japan credit cards are accepted only by the largest retailers.

FIRST UK CREDIT CARD

Barclays Bank launches Barclaycard, the first British credit card, which is also the first outside the USA. From the early 1960s, cards become more prevalent as competition between providers grows.

1966 XXXXXXX XXXX

CHIP AND PIN

Chip and PIN technology is trialled in Britain and is phased in from 2004, replacing payment authorization by signature.

2003 XXXXXXX XXXX

FIRST ATM MACHINE

The world's first cash dispenser or ATM (Automated Teller Machine) is installed by Barclays Bank in London. Before plastic bank cards were developed special cheques had to be used.

1967 XXXXXXX XXXX

FIRST UK DEBIT CARD

The first UK debit card, where the purchase is immediately charged to the holder's account, is launched by Barclays.

1985 XXXXXXX XXXX

Computer Games_1951

Ralph Baer (USA)

In 1951 German-born American engineer Ralph Baer discovered that data projected on to television screens could be manipulated by the viewer, creating an interactive environment ideal for games. His employers would not allow Baer to develop the idea and it was not until 1966 that he created Chase, the first video game to run on a standard television set. Meanwhile, Spacewar! the first game designed for computers, was devised in 1961 by students in Massachusetts. The principles are the same, with the gamer playing by inputting commands electronically. In computer games this is achieved using the mouse and keyboard, while in video games the player uses a joystick. These are played on a dedicated console rather than a PC.

Something to Think About ...

Most gamers prefer either video or computer games. Familiarity with one set of controls tends to preclude shifting between the two. Choose your weapons ...

1958
Tennis For Two is devised by American physicist William Higinbotham in New York. Another tennis-based game, Pong, is launched by Atari Inc. in 1972 achieving great success.

1952 A. S. Douglas develops a computer version of noughts and crosses at Cambridge University in the UK.

1966
Inventor Douglas Englebart pioneers the principles of virtual reality, a computer-generated environment with which the user interacts as if it were real. It is used for military training as well as gaming.

1972
Developed by Ralph Baer, the first video games console is released. Perhaps the most successful console, the Sony PlayStation, is introduced in 1994.

1995
Personal computer games are revolutionized by the introduction of the Graphical User Interface (GUI), which simplifies and improves the input of commands.

Barcode_1952

Bernard Silver & Norman Woodland (USA)

Bernard Silver and Norman Woodland were graduate students in Philadelphia in 1952 when they devised the barcode, a method of automatically reading product information at the checkout in grocery stores.

Barcodes assign items with unique 12-digit codes, represented by differing line widths. Using red light, a scanner reads the information and sends it to a computer. It was not until the 1970s that the system came into widespread use in America as the Universal Product Code (UPC). The technology was extended to most other retail products and adapted for a multitude of areas, including hospitals, airports and to instruct microwave ovens how to cook food.

1959

In the USA MIT graduate David Collins begins work on a barcode-recognition system for freight trains for Sylvania/GTE. In 1969 Collins leaves to start Computer Identics Corp., the first company whose product line is based entirely on barcodes.

Something to Think About . . .

Woodland and Silver were undoubtedly ahead of their time in inventing the barcode. It took two decades to catch on. Neither man became rich as a result of their creation, however, and Silver died in a car crash in 1963.

1970

The American National Association of Food Chains convene a committee to consider product identification and barcodes.

1972

In a supermarket in Ohio, a pack of chewing gum is the first barcoded product sold. The wrapper is now in the Smithsonian Museum, Washington.

1979

On 7 October 1979, the first barcode was used in the UK at Key Markets in Spalding, Lincolnshire.

1988

2-D barcodes, a pattern of small black and white squares on an area the size of a postage stamp, were introduced by Intermec for use on envelopes. 2-D barcodes now come in a variety of shapes, in circular designs and in colour.

Satellite_1957

Sputnik program (USSR)

The USSR's success in this pioneering first lap of the Space Race was not merely a political embarrassment to the USA. There were also fears about the military threat it represented – the rockets used to launch Sputnik were modified Intercontinental Ballistic Missiles. Today, satellites are a vital tool of military Command and Control infrastructure. GPS, the Global Positioning System, was originally developed for US military purposes, but released for civilian use in 1998.

Other civilian benefits came from Telstar, the world's first communications satellite, launched on 10 July 1962. It made long-distance telephone calls a reality and paved the way for satellite television. Despite improvements in ground cable technology and fibre optics, satellites still deliver phone and Internet services to remote areas.

Something to Think About . . .

Weather satellites monitor not only meteorological conditions but other environmental events. Man-made pollution, and natural disasters such as volcanic ash clouds, can have as great an impact on our domestic and working lives as wind and rain.

29 July 1955
The USA declares it will launch an artificial satellite as its flagship event during International Geophysical Year (IGY) 1957–8. Development is placed in the hands of the US Navy.

9 August 1955
The Soviet Union announces the Sputnik satellite program, also aimed at IGY.

15 May 1958
The Soviet Union achieves yet another successful satellite launch with Sputnik 3.

17 March 1958
The US Navy finally launches Vanguard 1. Solar powered, it transmits data for seven years, compared to Sputnik 1's three weeks. As of 2011 it is the oldest satellite still in orbit.

4 October 1957
The launch of Sputnik 1 catches the USA by surprise. Its cheery beep-beep signal, monitored by radio hams every 96 minutes, is a constant embarrassment.

1 February 1958
To the discomfort of the US Navy, the US Army launches its own satellite, Explorer 1.

3 November 1957
Sputnik 2 carries Laika the dog into space. She survives the launch but dies soon after.

6 December 1957
The US Navy's test rocket explodes on take-off, as does another on 5 February 1958.

Contraceptive Pill_ 1960

Carl Djerassi & Frank Colton (USA)

The first birth-control pill was developed by Austrian-American chemist Carl Djerassi in 1951. Two years later, Frank Colton synthesized Enovid, a chemical steroid that formed the basis of the first freely available oral contraceptive. In 1960, the United States Government gave approval for the Pill to be marketed. Most versions of the Pill contain two hormones, oestrogen and progestin; the mini pill contains only progestin. These are similar to natural female hormones and prevent the release of eggs so that fertilization and pregnancy cannot take place.

Something to Think About ...

The Pill had some prominent opponents. President Dwight Eisenhower stated in 1959 that birth control was not the business of government. After a Vatican committee reviewed the issue, hopes were high that the Catholic Church might approve the Pill, but these were dashed in 1968 when Pope Paul VI confirmed its strict anti-birth-control stance.

1961
The Pill (as it quickly becomes known) is made available via the National Health Service in Britain.

1963
In the USA 2.3 million women are taking the Pill, a factor that arguably fuels the 'Swinging Sixties' as women are able to control their own fertility.

1965
The United States Supreme Court strikes out the Connecticut state law banning the Pill. Massachusetts prohibits its use by unmarried couples until 1967 when the Supreme Court also declares this unlawful.

The birth-control revolution

1980
Safer, low-dosage contraceptive pills are introduced, the original high-dosage version being withdrawn from the market towards the end of the decade.

1967
Worldwide, 12.5 million women are using the Pill. By 2011 this figure will have increased to more than 100 million.

Laser_ 1960

Theodore H. Maiman (USA)

The first functioning laser was operated by Theodore H. Maiman in California in 1960. Maiman, working for Hughes Research Laboratories, won a race between several organizations to build the first laser, the principles of which had been laid down by Albert Einstein in 1917.

Laser is an acronym for Light Amplification by the Stimulated Emission of Radiation. It concentrates light over and over again until it emerges in a very powerful beam, which, depending on its strength, can cut through metal or be projected high into the sky. Lasers are now used in CD and DVD players, in manufacturing and in medicine, most commonly for corrective eye surgery.

1953
American physicist Charles Townes develops a 'maser', working on similar principles to the laser but using microwave radiation instead of light.

1959
In the USA another physicist, Gordon Gould, uses the acronym LASER for the first time in a research paper.

1960
Closely following Maiman's invention, Ali Javan and William R. Bennett Jr. demonstrate the first gas laser.

Something to Think About . . .

Lasers were anticipated by science-fiction writers going as far back as 1898 when H. G. Wells' *War of the Worlds* featured a Heat Ray weapon that destroyed everything in its path. Similarly, the 1930s comic strip *Buck Rogers* featured a disintegrator ray.

1962
While working at General Electric in New York, engineer Robert Hall develops a new type of laser, the semiconductor, which is now used in electronic appliances and communications systems.

1974
Lasers are used in everyday life for the first time in supermarket barcode scanners.

Chapter 08.0

Manned Space Flight to the iPad and beyond 1961 —

Manned Space Flight_1961

Yuri Gagarin (USSR)

If the USSR led the early stages of the Space Race with the first satellite (1957) and the first man in space (1961), the USA caught up and overtook it with the Apollo Moon landings (1969–71). The race was formally ended in 1975 with the joint Apollo-Soyuz space flight. Construction of its successor the International Space Station began in 1998.

When flights in the reusable Space Shuttle began in 1981, space travel became almost routine, a less romantic, more commercial affair. The exploration of deep space has been left to unmanned probes – so far. But closer to home, space tourism has become a reality if you can afford the $20 million price tag for a trip to the International Space Station.

12 April 1961

Soviet cosmonaut Yuri Gagarin is the first man in space, circling the Earth once aboard Vostok 1. On 16 June 1963 the USSR scores the double when Vostok 6 orbits for three days carrying Valentina Tereshkova, the first woman in space.

5 May 1961

Alan Shepard is the first US astronaut in space in a 15-minute suborbital flight on the spacecraft Freedom 7. Twenty days later President Kennedy launches the Apollo space project, declaring that America will put a man on the Moon by 1970.

Something to Think About . . .

Only 12 men (and no women) have walked on the Moon so far, all American. NASA's plans to return to the Moon in 2020 have been cancelled, but Russia is developing a new reusable space vehicle for manned Moon missions. India has projects for manned interplanetary flight and Europe has a long-term plan for a manned mission to Mars.

20 July 1969

Neil Armstrong and Buzz Aldrin are the first men on the Moon, with Michael Collins in the command module, four days after leaving the Earth in Apollo 11. It is the first of six successful US lunar missions over the next 30 months. The USA is the only nation to have landed on the Moon.

The race to the Moon

Microprocessor_1971

**Frederico Faggin, Ted Hoff
& Masatoshi Shima (USA/Japan)**

In 1965 Gordon Moore, co-founder of microprocessor giant Intel, formulated Moore's Law. The Law states that the number of components possible in an integrated circuit doubles every two years, a reflection of the rapid pace of progress in the electronics industry.

Jack Kilby of Texas Instruments developed the first microchip in 1958, an integrated circuit in which all the components are made from the same material, significantly reducing not only its size but also the time and cost of manufacture and assembly. Before microchips, every element of an electronic circuit had to be individually assembled on circuit boards, wired up and soldered.

When Intel was approached by Japanese electronics firm Busicom to supply the chips for their new Busicom 141-PF calculator, Intel took the opportunity to push the boundaries of microchip capability and function. In 1971 they reduced the circuitry of the calculator to four main microchips – together, effectively, the workings of an entire computer. The fourth, named 4004, was the first ever mass-produced CPU chip, or microprocessor.

The CPU (Central Processing Unit) is the brain behind all modern computers, and by producing one on a single chip, Intel transformed the computer age. Now microprocessors are the vital parts of everything from kitchen equipment to cars, from computers to credit cards.

Something to Think About . . .

In 1946 the US Army developed ENIAC, the first general-purpose computer. It covered 167 sq. m (1,800 sq. ft), weighed 30 tonnes and contained five million hand-soldered connections. In 1971 the Intel 4004 microprocessor, the size of a grain of rice, was more powerful than ENIAC.

2,300

The number of transistors built into Intel's 4004 microprocessor in November 1971.

3,500

The number of transistors in the 4004's successor, the Intel 8008, in April 1972.

2 billion

The number of transistors incorporated in the circuitry of Intel's 2008 Tukwila chip.

3.1 billion

The proposed number of transistors in Intel's Poulson chip due for release in 2012.

Internet_ 1973

Vint Cerf & Bob Kahn (USA)

The Internet is a global network of computers, the threads of a web. Although open to all who are connected to it, it was until 1991 a technically challenging system for those untrained in computer programming. The tools invented by Tim Berners-Lee (HTML, URL, HTTP) gave ordinary Internet users a simpler language, a system of addresses and a set of rules for the interconnection of web pages – the foundation of the World Wide Web. The World Wide Web is effectively an application for using the Internet.

The Web is managed by the World Wide Web Consortium, of which Berners-Lee is now Director. ARPANET, the first computer network, became eventually just one of many networks linked together as the Internet, which is now regulated by the Voluntary Internet Engineering Task Force. ARPANET itself was wound up in 1990.

Something to Think About ...

Computers, the Internet, and the World Wide Web have come a long way in a very short time. But we can all sympathize with the experience of the student programmer who tried to transmit the very first Internet message on ARPANET on 29 October 1969. Attempting to send the word 'LOGIN', he managed the L and the O; but as he typed the G, the whole system crashed. Still, 'Lo!' seems an appropriate first word for the invention, which in less than 50 years has changed almost every aspect of human activity.

1972

For a much-expanded ARPANET, American programmer Ray Tomlinson devises a system of electronic mail. The first e-mail ever sent is the message 'QWERTYUIOP'.

1969

ARPANET, brainchild of Joseph Licklider at America's Advanced Research Projects Agency, links four computers at UCLA, mainly to conduct more powerful calculations.

1973

Computer scientists Vint Cerf and Bob Kahn begin to develop TCP (Transmission Control Protocol) and IP (Internet Protocol), making this the first use of the word Internet.

1989–91

English programmer Tim Berners-Lee devises HTML (Hypertext Markup Language), URL (Uniform Resource Locator) and HTTP (Hypertext Transfer Protocol).

1983

Cerf and Kahn's TCP and IP, which allow any computer to use ARPANET, become the standard rules for Internet access; ARPANET is opened up to the public.

Mobile Phone_1973

Martin Cooper (USA)

Mobile phones have come a long way since the first car phones, which appeared in Sweden in 1960 and weighed 40kg (88lb). The first call from a genuinely portable phone was made by its inventor Martin Cooper of Motorola to his rival at Bell Labs, Joel Engel, on 3 April 1973. By 2011 Cooper had been joined by 5.3 billion other mobile-phone subscribers, over three-quarters of the world's population.

The first short message service (SMS) was launched in 1994; the first downloads, personalized ringtones, became available in 1998; and in 1999 the first payment facilities and full Internet access via mobile phone. 3G now delivers content not only to phones but to netbooks and eReaders, and our mobile phones are becoming smaller and smaller while incorporating infinitely more functions than the mere art of conversation. For many of us, our phones contain our whole lives of work, leisure and social activity.

oG

Early wireless phone networks are limited in coverage to one base station, which can serve a very small number of simultaneous users – as few as 23 in early technologies.

Something to Think About ...

In the 1980s a report for AT&T suggested that mobile phone ownership would reach 900,000 by the end of the millennium. That was before the introduction of 2G technology. In the early years of the 21st century it is estimated that 900,000 mobile phones were being sold every three days. The total market figure in 2000 was closer to 600 million.

1G

For the first time calls can be continued as the user moves between stations. The first service is launched in 1979 by NTT, covering all of Tokyo with a network of 23 cells.

2G

A move from analog to digital signal begins in 1991 with Radiolinja's GSM network in Finland. Mobile phone use expands; compact handsets replace the old 'bricks'.

3G

Anticipating the use of mobile phones for more and faster data transmission, Japan's NTT network tests 3G in 2001. Internet access via the phone becomes commonplace.

4G

Traditional phone technology is abandoned in favour of Internet protocols, with speeds 100 times faster than 3G, treating voice transmission just like any other data.

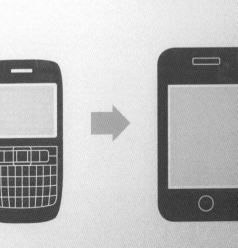

Personal Computer_1974

Ed Roberts & Bill Yates (USA)

Until 1974 computers were enormous, expensive pieces of equipment only affordable by larger businesses and corporations. In that year American electronics enthusiasts Ed Roberts and Bill Yates launched the Altair 8800 minicomputer, which was small enough to fit on a desk. It was sold in kit form and promoted in *Popular Electronics* magazine's January 1975 issue. In July that year they introduced a programming language called Altair BASIC, the first ever product from a new company called Microsoft formed by a young Paul Allen and Bill Gates.

Within two years the Altair had spawned many imitators including, in 1976, the first Apple computer. The rivalry between successive operating systems developed by Apple and Microsoft continues to this day, and their efforts to dominate the personal computer market have not always been to the ultimate benefit of the consumer. But the arrival of small computers transformed first the business world, then our home and school life, then – through social networking – the nature of society itself.

Something to Think About ...

The Commodore 64, launched in 1981, did more than any other early PC to bring computing into people's homes. Adopting mass production techniques, and marketing itself through retail stores rather than specialist electronics outlets, the C64 sold up to 17 million units. Its success in turn led developers to write over 10,000 software titles for the machine, demonstrating to those early domestic users just how useful computers could be in every aspect of our daily lives.

**** PERSONAL COMPUTER BASIC V2 ****

READY. ■

INFORMATION IN

Early pre-desktop personal computers (PCs) rely on the
familiar typewriter keyboard to input data. American
computer pioneer Douglas Engelbart devises a wheel-based
mouse in 1963, and German firm Telefunken a ball-driven
one in 1968. The mouse takes off in 1984 when Apple designs
a Graphic User Interface for their Macintosh. This allows
ordinary users to point and click instead of typing in
complex computer-language commands. Touchpads first appear
in 1983.

INFORMATION OUT

Home PCs naturally adopt the cathode-ray tube (CRT)
technology of the domestic TV set. Monitor displays are
monochrome until IBM introduces colour graphics adapters
in 1981. Lightweight energy-efficient liquid crystal
display (LCD) screens replace CRTs in the early 2000s.
In the future, light-emitting diode (LED) monitors will
offer better viewing angles and higher contrast.

INFORMATION ON THE MOVE

The move from desktop to laptop begins in 1981 with the
Australian Dulmont Magnum model. 1983's Gavilan SC is
first to use the term laptop and to include a touchpad.
From laptop to handheld, Psion launch their functionally
and physically compact Organiser in 1984. Asus splits the
difference with its EeePC Netbook in 2007.

08.6 Sticky Notes_1974

Arthur Fry & Spencer Silver (USA)

They come in all shapes, sizes and colours, the perfect solution to the problem of how to leave temporary messages and markers without leaving permanent marks by the telephone, in a book, or on the map.

It's a mark of the sticky note's success as an office tool that even paperless computer applications have adopted the familiar form of yellow squares, for users to leave themselves on-screen notes.

Some inventions are driven by urgent necessity; some inventors are driven to constant experimentation; but some ideas are just accidental discoveries. Nobody went looking for sticky notes. But when the idea presented itself, we immediately wondered how we ever managed without them.

1970

Spencer Silver, a research chemist in the labs at 3M, the American conglomerate based in Minnesota (formerly known as the Minnesota Mining and Manufacturing Company), invents a very weak glue while attempting to invent a very strong one.

'What use is that to anyone?'

he wonders.

Something to Think About ...

The yellow paper most closely associated with the product was yet another happy coincidence in the history of its development – there just happened to be some yellow paper lying around in the 3M lab next to the one in which Arthur Fry was experimenting with Spencer Silver's glue.

1973

3M product development researcher Arthur Fry is frustrated by the bookmarks that keep falling out of his hymnal in the church choir, which leads him to ponder,

'What would keep them in place?'

1974

Fry experiments with Silver's glue, strong enough to hold bookmarks to the page, but weak enough to remove without marking.

'Can I market this?'

he asks his boss.

The answers to some sticky questions

1977

3M begins a trial launch of the product, called 'Press'n'Peel', in four American cities. Sales are discouraging but they like the idea in Boise, Idaho. They ask themselves the question,

'Do we need to rebrand?'

1980

Post-It Notes, the original sticky notes, are launched across the USA, and a year later in Canada and Europe. They are an instant success.

No more sticky questions.

Walkman_1979

Nobutoshi Kihara (Japan)

Sony engineer Nobutoshi Kihara built the first Walkman in 1978 so that the founder of Sony, Akio Morita, could listen to music during his frequent plane trips. The design was refined using microelectronics combined with the cassette format to create a smaller, lighter player equipped with compact headphones. The Walkman was launched in Japan in 1979. It created the market for personal stereos and changed the way in which music was listened to by making it truly portable for the first time. The Walkman went on to sell more than 220 million units in over 300 different models.

1983
The cassette-based Walkman is the latest craze worldwide, its name already synonymous with personal stereos.

1984
Sony produces the CD (Compact Disc) Walkman, known initially as the Discman. Later models are improved by anti-skipping technology. During the same year, a recording Walkman is unveiled.

1985
In the face of competition from rival products, the Walkman is refined to feature graphic equalizers (sophisticated tone controls) and improved sound quality.

Something to Think About . . .

The name 'Walkman' came about because there is no direct translation of 'transportable' in Japanese, the nearest equivalent being 'walking man'. Sony chairman Akio Morita disliked the name and the device was originally marketed as the Soundabout in America and the Stowaway in Britain.

1989

The Video Walkman is Sony's latest innovation, using tapes in the small Video-8 format common to camcorders.

1999

With the cassette format quickly becoming obsolete, Sony enters the digital audio market with a new version of the Walkman that stores music on a memory stick in MP3 form.

Music on the move

08.8 Compact Disc_1979

Philips/Sony Compact Disc Taskforce (The Netherlands)

Electronics company Philips first demonstrated the Compact Disc at a 1979 conference, and then established a taskforce with Sony to develop it. The new medium used digital optical recording with playback via a laser. CD players were launched worldwide in 1983, along with a selection of discs. As the price of hardware fell and the range of titles increased, CDs gained popularity and, by 1988, more than 400 million were being pressed annually. The format would later be used for data storage and software applications.

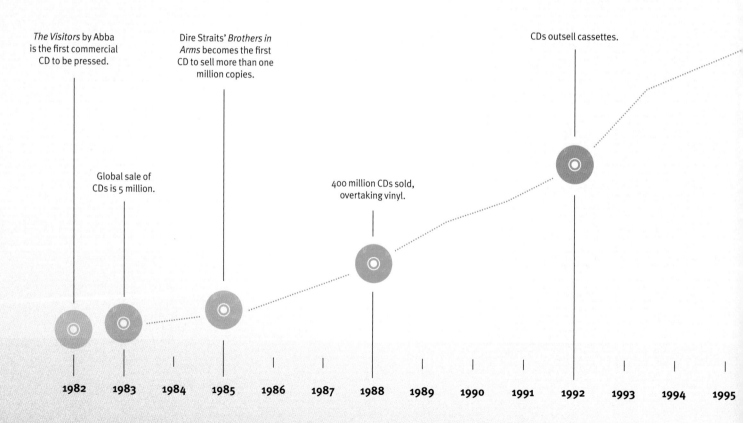

The Visitors by Abba is the first commercial CD to be pressed.

Dire Straits' *Brothers in Arms* becomes the first CD to sell more than one million copies.

CDs outsell cassettes.

Global sale of CDs is 5 million.

400 million CDs sold, overtaking vinyl.

1982 1983 1984 1985 1986 1987 1988 1989 1990 1991 1992 1993 1994 1995

Something to Think About ...

The maximum CD playing time was originally to be 60 minutes. It was extended to 74 at the request of Sony executive Norio Ohga to accommodate Wilhelm Furtwängler's recording of Beethoven's Ninth Symphony.

First DVDs

In 1995, Philips, Sony, Toshiba and Panasonic adapted CD technology to create the Digital Versatile Disc (DVD), which is the same size as a CD but has enhanced storage and multimedia capacity. The format was introduced in Japan in November 1996, in the USA in March 1997 and in Europe in October 1998. By 2003, DVD had outstripped video in America. A potential successor, Blu-ray, arrived in 2006.

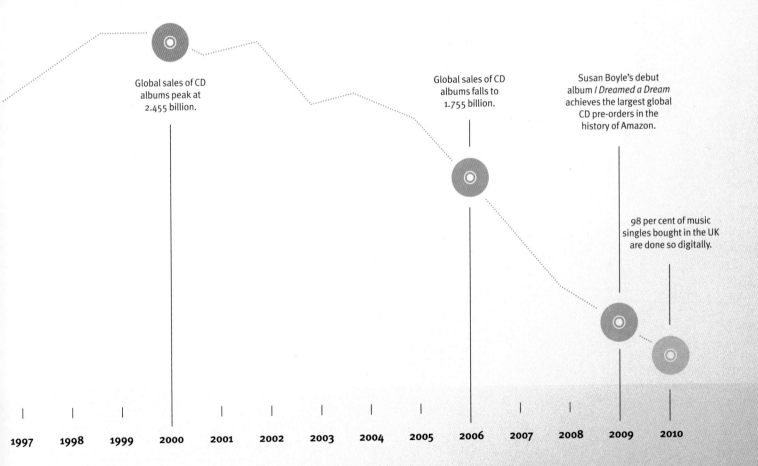

Global sales of CD albums peak at 2.455 billion.

Global sales of CD albums falls to 1.755 billion.

Susan Boyle's debut album *I Dreamed a Dream* achieves the largest global CD pre-orders in the history of Amazon.

98 per cent of music singles bought in the UK are done so digitally.

1997 1998 1999 2000 2001 2002 2003 2004 2005 2006 2007 2008 2009 2010

DNA Fingerprinting_ 1984

Sir Alec Jeffreys (England)

While studying the results of an experiment in 1984, geneticist Alec
Jeffreys was struck by the realization that variations in deoxyribonucleic
acid (DNA) could be used to identify individuals. DNA is a genetic
code found in every human cell. It contains the instructions for the
development and functioning of the body. Certain repeating patterns in
the chemical structure are unique to individuals and two DNA samples
can be compared to discover whether they are from the same person,
from blood relatives or from unrelated people. The process became
known as DNA fingerprinting.

Something to Think About . . .

DNA fingerprinting prevented a
miscarriage of justice in the UK in
1988 – the very first time it was used
in a criminal trial. Without DNA
evidence, it is likely that the prime
suspect, Richard Buckland, would have
been convicted of the murder and rape
of two teenagers committed by
Colin Pitchfork.

1984
Working with the UK Forensic
Science Service, Jeffreys
starts to develop DNA
profiling by refining his work
to make it more sensitive and
computer-friendly.

1985
The first use of DNA fingerprinting in the world is in a British immigration case where the tests determine a mother-son relationship.

1992
German prosecutors verify the identity of Nazi war criminal Dr Josef Mengele by DNA testing his exhumed elbow bone and comparing it to samples from other family members.

1988
In Britain DNA evidence is used for the first time in a criminal trial.

1987
Jeffreys' techniques are commercialized, allowing them to be used throughout the world in criminal investigations and paternity testing.

MP3_1996

Karlheinz Brandenburg, Bernhard Grill, Thomas Sporer, Bernd Kurten & Ernst Eberlein (Germany)

The MP3 was developed by a team of researchers in Germany led by Karlheinz Brandenburg and was patented in America in 1996. Their goal was to devise an audio encoding format using data compression techniques that had not previously been applied to sound. By removing redundant information inaudible to the human ear, the amount of data required by the original recording is significantly reduced without appreciable loss of listening quality. Conversion to MP3 reduces the size of a CD recording by between 10 and 14 times. MP3s makes downloading faster and allow large amounts of music to be stored on the hard drive of a personal player.

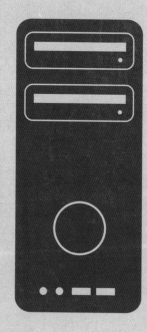

Something to Think About . . .

The development team favoured simply recorded music when testing the MP3 because faults were easier to detect. Suzanne Vega's 'Tom's Diner' was the first song transferred to the new format, leading the singer to be dubbed 'the mother of MP3'.

Music becomes invisible

Following vinyl, cassette and CD, the MP3 offers music in a digital format that requires no physical release.

2002
The Archos Multimedia Jukebox is the first MP3 multimedia player.

1996
American technology entrepreneur Nathan Schulhof's company Audio Highway produces the first portable MP3 player, the award-winning Listen Up, in a limited edition of 25.

2001
Apple introduces the first iPod. Initially both this and early digital Walkman models operate on the companies' rival encoding formats.

1998
Early MP3 players have limited storage capacity until Compaq launches the first model with a hard drive.

1999
With piracy and illegal copying of music via MP3 a controversial issue, American record label Sub Pop becomes the first officially to issue music in MP3 format.

08.11 iPad_2010

Apple Inc. (USA)

The iPad is a high-profile example of the tablet computer, the most recent development in personal computing. It illustrates the debt owed by inventors of any age to the pioneers who went before them. The iPad could not exist without the earlier development of microprocessors, transistors and indeed the electric cell itself; without mankind's exploitation of materials from specialized glass to plastics; without the ground-breaking invention of communications systems such as the telephone, the telegraph and the wireless radio. At least a third of the contents of this book have paved the way for the tablet computer.

Something to Think About . . .

Graphene, a new material, is set to revolutionize the world of electronics. It is exceptionally strong, light and conductive. Samsung have already demonstrated its use in a 25-in (63.5-cm) flexible touchscreen, raising the possibility of a tablet the size of a credit card or an iPad that you can roll up for compact storage.

November 2001

Computer giant Apple's digital music player, the iPod, is launched – intended as a stylish, compact rival to existing MP3 digital audio players, although it still can't (or won't) play MIDI or Windows files. By mid-2011 we will have bought over 307 million iPods, including 60 million of the iPod Touch.

2001

April 2010

Since the iPhone is already much more than just a phone, Apple's arrival in the tablet computer arena with the iPad is no surprise. After one year, it already has a 75 per cent market share, having sold nearly 15 million units worldwide.

June 2007

Apple's entry into the smartphone market, the iPhone is the first to use multi-touch technology – screen control for more than one finger. It gives us a new word: *apps* (short for applications). By 2011 almost 400,000 licensed software applications for the phone will have been created.

08.12 What's next?

As we come closer to a full scientific understanding of the world we live in, we now know (more or less) the basic laws of nature. With little left to discover, scientists and inventors must now concentrate on applying those laws in innovative, responsible ways. Norwegian Nobel Physics Prize winner Ivar Giaever said in a speech in 2008 that 'the number of scientific laws is finite, but potential inventions are limitless'.

So what next? What are the needs, the necessities that will drive the next generation of inventions? Will we all be teleporting to our holiday destinations, or will future vacations be virtual experiences? Will we all be hard-wired to the Internet? Or will the Internet be a thing of the past?

The searches for more powerful, more efficient ways of doing things will inspire inventors as they have always done. But new challenges are presented by the scarcity and pollution of the Earth's resources, such as its water and fossil fuels. We can expect inventions for saving, purifying and protecting water, as well as inventions that will offer cleaner,

iPod (2001)

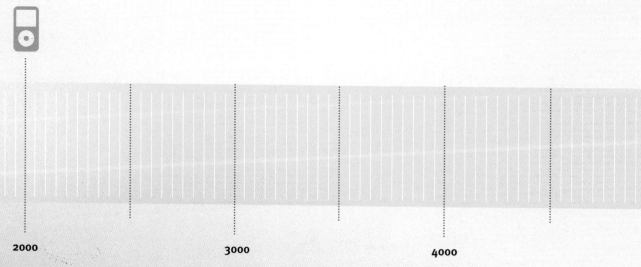

2000   3000   4000

sustainable alternatives for our ever-increasing demand for energy. Elsewhere, inventions will take advantage of new materials such as graphene, which is lighter and stronger than silicon, in the continuing trend for miniaturization.

Genetically modified plants and animals are already with us and we are beginning to understand the genetic causes of many human conditions. Many future inventions may be driven by biological science. Cosmetic surgery is commonplace, but there are moral concerns about the extent to which science should be used to adapt the human body both superficially and genetically.

Throughout the ages, the basic human desire to communicate has thrown up many inventions. Whatever new media we invent to serve that urge, the great thing is that we will continue to exchange views and share ideas. On those shared ideas will be built new ideas, the inventions of the future.

Index_

Index_ *cont.*

Authors

Michael Heatley is the author of over one hundred books, ranging in subject matter from music biography and popular culture to sports and general reference. He is the author of *D is for Dad: Survival Tips for the Modern Father* and *The Dads' Book: For the Dad Who's Best at Everything*.

Colin Salter is a history and science writer with a fascination for how things work, and how they used to work. He has written about everything from the sinking of the *Titanic* to the private lives of marine gastropods. His contributions to *Chambers' Biographical Dictionary* include the entries for 500 living scientists. This is his seventh collaboration with Michael Heatley.

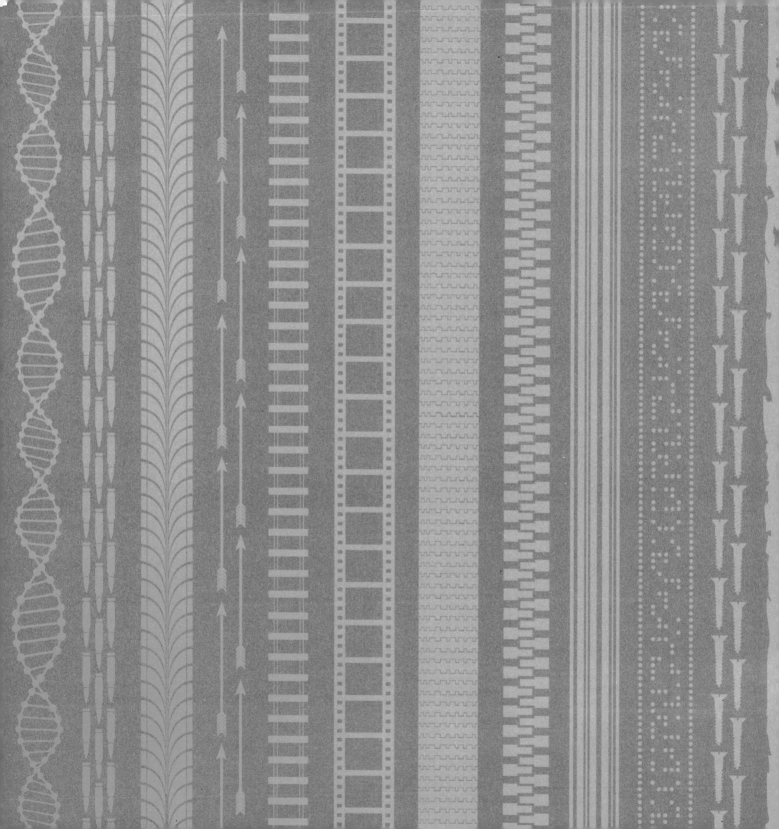